Vocational Guide to Modern Television Systems Servicing

Vocational Guide to Modern Television Systems Servicing

Rajneesh Goel

RANDOM PUBLICATIONS
NEW DELHI (INDIA)

Vocational Guide to Modern Television Systems Servicing

ISBN 978-93-5111-406-2

Published in 2014 in India by

RANDOM PUBLICATIONS

4376-A/4B, Gali Murari Lal, Ansari Road
New Delhi-110 002
Phone : +91-11-43580356, +91-11-23289044
e-mail: randomexports@gmail.com, sales@randompublications.com, info@randompublications.com

Type Setting by : Keystoneprintads, Delhi-110051
Printed at Thomson Press (India) Ltd

Preface

Television, like any modern invention of complexity, did not simply appear-it evolved. Much of its technology was and still is based on previous inventions as well as methods used by radio. It is not within the scope of this paper to present the history of the telegraph, telephone and radio-suffice it to say that each of these technologies built upon its predecessor.

All but one analog television system began as black-and-white systems. Each country, faced with local political, technical, and economic issues, adopted a color television system which was grafted onto an existing monochrome system, using gaps in the video spectrum (explained below) to allow color transmission information to fit in the existing channels allotted. The grafting of the color transmission standards onto existing monochrome systems permitted existing monochrome television receivers predating the changeover to color television to continue to be operated as monochrome television. Because of this compatibility requirement, color standards added a second signal to the basic monochrome signal, which carries the color information.

The color information is called chrominance with the symbol C, while the black and white information is called the luminance with the symbol Y. Monochrome television receivers only display the luminance, while color receivers process both signals. Though in theory any monochrome system could be adopted to a color system, in practice some of the original monochrome systems proved impractical to adapt to color and were abandoned when the switch to color broadcasting was made.

Television bookmarks our lives, preserves moments in time that trigger memories of where we were and what we were doing when historic events occurred. Its images and sounds touch our emotions, connect us with world events, and help us prepare for natural disasters with weather reports using equipment so sensitive, it tracks storm moving block by block. We are entertained, informed, influenced and motivated to action by TV. As an industry, television is an exciting business that continues to thrill. Look around when a camera crew arrives on a scene and the lights come on. Even the most conservative people may be swayed by the allure of TV and clamor to be in that spotlight. This book is prepared for journalists, news readers, researchers, publishers, students and for all those associated with media, broadcasting and publishing industry.

I thank all members of my team who have helped in the preparation of the book. My special thanks go to "Random Publications" who have published the book.

– ***Rajneesh Goel***

Contents

1

Career in Television

WHY CONSIDER A CAREER IN TELEVISION

Television bookmarks our lives, preserves moments in time that trigger memories of where we were and what we were doing when historic events occurred. Its images and sounds touch our emotions, connect us with world events, and help us prepare for natural disasters with weather reports using equipment so sensitive, it tracks storm movement block by block. We are entertained, informed, influenced and motivated to action by TV.

As an industry, television is an exciting business that continues to thrill. Look around when a camera crew arrives on a scene and the lights come on. Even the most conservative people may be swayed by the allure of TV and clamor to be in that spotlight.

STARTING YOUR CAREER

"Think big, start small."

That's the advice given to industry newcomers by vice presidents and general managers of TV stations interviewed, especially to those seeking work in news.

Bigger markets, for example, won't hire people in a number of positions who lack specific television or related industry experience. Says one president/ GM with a large Hearst-Argyle station, about hiring novices in the newsroom: "[They] can be costly if they do not know the rules of journalism and basic laws of libel and slander. We prefer our talent to have prior experience before they come here."

There are exceptions. One station head personally recruited a broadcast student at a local university known for its outstanding journalism program who was a prior station intern, graduated with a 3.9 average, came highly recommended by her professors and was named an outstanding student in her field by the college. During her internship, she impressed the GM and news director, and the station hired her when the news assistant went out on maternity leave. Their satellite bureau then needed a production coordinator and she's now a full-time staffer there.

How much experience you will need and the education level required depends on the station you are interested in and the job sought. Says one president/GM: "I actually prefer our talent to have more of a liberal arts background than to have focused exclusively on broadcast journalism, because what you learn in school is quite different from the realities of the TV station. Those with broader knowledge bases are apt to be more versatile in their reporting."

STARTING YOUR TELEVISION CAREER

Most of the Hollywood studios run their own writing programs aimed at training new writers for their shows. That of Warner Bros is one of most well-established, having started in 1976. Others include Disney's Chesterfield Fellowship and Fox's Diversity Program. Starting Your Television Career is intended as a companion title to aid newcomers to the course but is of great use to other novices.

After an initial introduction, the book is helpfully divided into two sections, one looking at writing sitcoms, the other the one-hour drama - as the program itself is divided. Every year, 30 people are chosen to be interviewed for each program and from this number, 13 are invited to participate in either category. (It costs $30 to submit an application and the workshop fee itself is $495.) Applications come from all over the world even, in one instance, from Afghanistan. Those who complete the program are virtually guaranteed work in the industry as top agents, TV execs and producers seek out participants. The authors Abby Finer and Deborah Pearlman are both TV writers themselves and lecturers on the program.

The book takes the writer from the initial conception of a storyline to rewriting the first draft and is aimed primarily at writing a "spec" script - in other words, writing for an existing show. Traditionally, new writers would have to prove that they could successfully replicate the characters and tone of a show already on the air before being asked to submit their own ideas, though this is now beginning to change.

The book goes into detail into how to analyse a show into its constituent parts. Where, for example, do act breaks fall? How many scenes are there and how long are they? What kind of stories are told? This section alone is worth the cover price.

The final chapter, meanwhile, includes interviews with working writers and tutors on the program, where they reveal how they got into the business and helpfully what they look for in a sample script. Plus there's a helpful glossary of industry terms so that you can sound like you know what you're talking about in meetings. The book also gives examples of sitcom and drama outlines, as well as the scripts themselves - here the Hispanic sitcom, The George Lopez Show and the procedural drama Without A Trace. It's interesting to see in both cases how the final script differs from the original outline. Clearly written and easy to follow, Starting Your Television Career is an excellent starting point for the novice. It's an invaluable guide that reveals why US networks have had such success in the TV market and why US shows are sold the world over. Ron Hart took the program himself in 1996 and is a producer on According To Jim. He says: "If you break into this field and you find yourself sitting in a room with ten really funny people who are trying to make each other laugh, don't mess it up. After all, it's the best job in the world."

A WORD ABOUT INTERNSHIPS

Whatever the field of study, get an internship. As one station CEO notes: "Internships are the best way to gain practical, real-world experience and they are the perfect test model for station management to assess the intern's level of intelligence, enthusiasm and growth progress during their internship stint...If interns impress us, but are just not ready to work here, we'll guide them to work at stations in smaller markets, ask them to keep in touch and remind them to be patient."

Theory is fine. Familiarity with industry terminology is great. But there is no substitute for the real experience and sometimes the most practical aspects of getting hired are never covered in class. Says one technical news operations manager who can't stand wrinkled or incomplete resumes: "You go four or five years to school and they don't even teach you how to write a resume, put it in a nice folder so it will stand out....a person doesn't deserve a job if he sends in his resume with no phone number."

David Zurawik, the National Media Writer for The Baltimore Sun, has extensive experience covering the industry, is a radio pundit on media issues and also teaches a journalism course at Goucher College. He's taught elsewhere, interviewed journalism department heads and shares a sentiment with many who actually work in the media: students often are woefully unprepared for the real media world. The best schools will have teachers who worked or still work in the industry, have contacts and can help students get jobs. More importantly, they offer broadcast internships that hands down, are the best ways to gain industry entry. These more often than not are credit-only, unpaid programs, so if you seek one during the summer, be prepared to take on another, part-time or full-time paying job.

TV REPAIR EDUCATION AND TRAINING REQUIREMENTS

TV repairmen adjust, install and repair television sets and other home electronic entertainment devices. In addition to completing a vocational or community college program, prospective TV repairmen typically undertake an apprenticeship, internship or on-the-job training.

TV REPAIRMAN EDUCATION REQUIREMENTS

Education for aspiring TV repairmen typically comes in the form of certificate or associate's degree programs. These TV repair programs might be offered at trade schools, community colleges or vocational schools. While it is not necessary for employment, potential TV repairmen may pursue a bachelor's degree in electronics if they're interested in obtaining additional education. To enter into these programs, a high school diploma or GED is required.

Classes in TV Repair

Courses in a TV repair education program start with the fundamentals of television's form and function and then tackle more advanced topics. Advanced courses might focus on techniques using TV testing equipment, which is a vital component to troubleshooting these devices to perform repair and installation.

Toward the end of these programs, classes focus on advanced theory for troubleshooting, repairing, analyzing and servicing televisions. Performing these duties involves circuit work and voltage examination, along with understanding special effects and features of modern television sets. In some cases, advanced TV repair classes cover complex topics like luminescence and chroma, television and radio frequencies, transistors and amplifiers, signal stages, circuit analysis and monitor principles.

Training Requirements

TV repair education needs to be supplemented with actual work and training experience. Several paths are available to obtain this TV repair training, including internships or externships while still enrolled in a TV repair program. Occasionally, employers hire TV repairmen and provide on-the-job training.

Apprenticeship programs are an effective choice for acquiring TV repair training. The U.S. Department of Labor offers a list of state apprenticeship programs. TV repairmen work underneath a mentor in these apprenticeship programs and learn the skills necessary for this occupation directly from them. Training length varies, and TV repairmen must keep current on changing technology and repairing techniques.

BECOME A TV REPAIRMAN: EDUCATION AND CAREER ROADMAP

Find out how to become a TV repairman. Research the education and training requirements and learn about the experience you need to advance your career in TV repair.

Requirements for a TV Repairman

Since television sets are a featured appliance in many homes, the need for trained specialists who can diagnose and repair TVs is strong. As the price and complexity of television sets increase, many consumers may choose to repair rather than replace their set if a problem arises. A TV repairman will diagnose problems with televisions, replace worn or defective parts, and show customers how to operate their televisions and fine-tune their picture.

Though some TV repairmen may learn on-the-job, many employers require applicants who have completed some type of postsecondary program. Certification is also becoming increasingly important. The following table contains the main requirements outlined by the U.S. Bureau of Labor Statistics for TV repairmen:

	Common Requirements
Degree Level	Many, though not all, employers may prefer to hire TV repairmen who have completed some postsecondary education
Degree Field	Electronics or related field
Certification	Voluntary certification is available
Experience	No experience is necessary for entry-level positions
Key Skills	TV repairmen should have strong customer service, problem-solving and record-keeping skills; manual dexterity and an excellent sense of vision are also essential
Technical Skills	Expertise with diagnostic equipment, such as multimeters, voltmeters and oscilloscopes, may be helpful
Additional Requirements	TV repairmen should exercise caution while performing their work; this profession has a higher than average risk for injuries, such as shocks and burns

Complete Education Requirements

The minimum level of education required to become a television repairman is a high school diploma. However, many employers prefer to hire candidates who have completed some form of postsecondary education. For increased career opportunities, candidates could pursue a diploma, certificate or associate's degree in electronics repair or technology.

These programs are designed to prepare students for a wide range of technical issues relating to the creation and maintenance of electronic devices. Some of the subjects that an education program in electronics may cover include direct current circuits, solid-state devices, electronics technology, soldering technology, linear integrated circuits, wireless communications and microprocessors.

A student's coursework also provides a strong foundation in communications, mathematics and computer skills.

Gain Experience

Experience may be obtained through an internship or a postsecondary program. Prospective TV repairmen can work and learn under the supervision of an experienced professional. Interning will offer individuals a chance to assist in repairs and begin communicating with customers.

Become Certified

Certification is not necessary in order to pursue television repair jobs. However, earning certification can help a repair technician secure better job opportunities, since it is becoming more important to employers. The Electronics Technicians Association (ETA) reports that more than 10,000 professionals earn certification through the group each year. Similarly, the International Society of Certified Electronics Technicians (ISCET) offers a number of exams for professionals of various experience levels and technical specialties. In order to earn these certifications, TV repairmen must meet certain prerequisites and successfully pass certification examinations. Individuals should research the specific requirements for each exam, since they vary by credential.

HOW TO SELECT A TELEVISION REPAIR SCHOOL

Students interested in a short program covering the foundations of electronics can enroll in certificate programs, which can be completed in a year or less. Diploma and associate degree programs can take up to two years to complete; however, they may offer more in-depth study of various electrical and computer systems which could benefit students with career goals beyond television repair. Electronics programs can be useful because companies may prefer employees with broad electronics repairknowledge concerning equipment such as home theater systems and DVD players. Some repair businesses cover many types of home entertainment devices rather than just TVs, and they may seek employees who can repair any of these devices. Television repair jobs can also involve working with several varieties and formats of television, so students may find it beneficial to enroll in programs where they can learn about plasma, LCD and other models.

Individuals interested in self-employment or repair company ownership might want to consider enrolling in programs that offer business management, finance and accounting classes. In many cases, television repair workers deal directly with the general public, so programs that offer costumer relations courses may be beneficial.

Students are recommended to look for schools with up-to-date on-campus training facilities so they can familiarize themselves with circuits, wiring and digital electronics. Programs that offer internship opportunities with local businesses could provide additional experience. Independent study or projects can allow students to work in more specialized areas of electronics and television repair.

TELEVISION REPAIR PROGRAM OVERVIEWS

Certificate in Electronics Technology

Certificate programs in electronics technology may be completed in as little as one semester. Basic electronics knowledge can be obtained through courses and laboratory experiences in circuits, semiconductors and transistors.

Diploma in Electronic Systems

Diploma programs in electronic systems may be completed in less than two years. Coursework includes basic studies in electronics such as AC and DC currents, circuits and wiring. Students may also have the opportunity to study more advanced concepts in electricity such as motor controls and microcomputers.

Associate Degree in Electronic Engineering Technology

Associate degree programs include courses in general education, basic electric concepts, advanced electrical systems and computer systems. Students begin to learn more about applications of electricity in computers, instruments and other technology. Independent projects and internship opportunities may be available at this level of study.

PLASMA AND LCD TV REPAIR: TWO TRAINING OPTIONS

If you are trying to learn electronics repair, particularly flat screen television repair which is this blog's main subject, but are still undecided as to how you are going to go about achieving it then let me tell you flat out (no pun intended) that there are only two routes that you need to choose from. It is a fork in the road that you need to decide on before you proceed with your education in this area.

You could either learn how to repair an LCD TV or a Plasma TV through:

- Offline means where you attend real TV repair schools or classes locally being held, or through
- Online Plasma and LCD TV repair course offerings

If you're the sort of person who chooses to learn things in a structured, four-cornered environment, with an real live instructor or teacher whom you can query and interact in real time and space, then you should go for a classroom course and attend one of those electronics classes being offered in vocational or electronic schools in your location. There are some nice advantages in taking this mode of learning. One is it affords a really timely and the immediate feedback mechanism when you go to the practical aspect of the course. Your instructor will be able to point out quite quickly whether you are doing it right or wrong. Another advantage is, since you have other students with you, you will also be able to pick up lessons from their experiences, most particularly their mistakes. Furthermore, you will be able to build up rapport with fellows that can serve as a technical support network that can come in handy in the future when you go into electronics repair full time (or even part time).

One great disadvantage of going to an actual repair school is that you will definitely get charged high training fees. After all, you will need to pay for the salary of the expert instructors and pay for resources and facilities that you will need to use. Another disadvantage of a classroom setting is that you may not have enough time to cover all aspects of the technical aspects of electronics repair. What they usually teach you are the fundamentals and the important advanced principles but some stuff are usually left out because classroom education usually have time limitations.

On the other hand, if you are the type of person who chooses to keep his own time doing things and work at his own pace and at the convenience of his own home, then you should opt for an online plasma or LCD TV repair course. What you will need simply is a computer connected to the internet and you can be on your way to becoming an electronics repair specialist. Remember though that you will not have some of the perks that school courses have. For example, You will need to put up your own troubleshooting, diagnostics and repair tools. Likewise, you will also not have that instant feedback mechanism that I mentioned in the alternative mode of learning LCD and Plasma TV repair. Obviously this is because there is no teacher or instructor watching your every move. When you go the DIY way, you will need to be deliberate and careful with your actions so as not to injure yourself or damage your property.

The great thing about learning it all online by yourself is that this method will definitely cheaper than a classroom repair course. That is because you won't have to pay teachers and you won't have to pay for depreciation of the tools and equipment that you're going to use. There are no investors (school owners) to keep happy but yourself!

The online method basically entails downloading information from the internet. That is it! Plasma and LCD TV repair manuals and books will be made available for you to download along with step-by-step guides to help you repair plasma TVs and LCD TVs. These materials will usually contain more information than what can be taught in any classroom short courses.

Online courses also offer great technical support via email forums and bulletin boards. In summary, if you are the sort of person who likes to learn through classroom interaction real live instructors or teachers then you should go for an offline classroom repair course. If you like to work at your own pace, download information, follow manuals, e-books, interact with fellow learners through online means and such, then definitely you should take the online repair course. The relative advantages and disadvantages that I provided above should help you decide on which method to choose to master the craft of electronics troubleshooting and repair.

OVERVIEW OF THE TV REPAIR BUSINESS

Thinking of starting a TV repair home business? It's not the lucrative job that it once was, but you still can make the most of this opportunity, or possibly consider pursuing it part-time or adding other electronics besides TVs.

New flat-panel TVs are far more complex than old analog sets were, making repairs more costly and more difficult. As a result, it is often cheaper for consumers to buy a new set than to repair an old one. But if you've got a knack for electronics and the right training, you can launch a TV repair home business. Odds are you won't earn enough to make it a full-time endeavor, but you might be able to earn enough to make it worth doing on the side. There are still plenty of homes out there with TVs that continue to provide some value in the home that could be repaired or updated without having to replace them. Because the business has faded, it's getting harder and harder for homeowners to find TV repair specialists. And, repairing older sets is often a much greener option than replacing them.

TV repair isn't a job or home business for novices. To start with, you should have some training — either through a TV repair apprenticeship or vocational school program — before you open for business. Some online and at-home correspondence courses also can be found in TV and electronics repair, as can product instruction manuals, but nothing beats old-fashioned hands-on experience. You might be able to supplement your income by refurbishing the first generation of flat-panel sets (the warranties for many of which have already expired), doing subcontracting repair work for an electronics retailer, or by repairing other consumer electronics like computers.

Pros of a TV Repair Home Business:

- *The market is growing rapidly*: A third of U.S. households owned a high-definition television in early 2009, according to the Leichtman Research Group. That's twice as many as just two years earlier. The transition from analog to digital broadcasting will likely prompt more people to buy flat-panel TVs or seek help in getting their old analog TV sets to work after digital TV transmission is mandatory.
- Some TV repairs, such as replacing a faulty panel, can be handled on the spot in the customers' home.

- The job can be good source of supplemental income for an electronics-savvy person.

Cons of a TV Repair Home Business:

- Manufacturer and retailer service centers can often do the repairs cheaper (but not all TVs are still under warranty and some manufacturers have since disappeared).
- Rapidly evolving technology requires staying up to date,as well as having the necessary knowledge to maintain older TV sets.
- Flat-panel TV prices are falling, meaning many customers will opt to replace, not repair, their existing set. However, in a down economy, customers's may not be able to afford a brand new set.

What You Need to Get Started in a TV Repair Home Business:

- Ample, secure space to repair TV sets and store them while they're undergoing repairs.
- A business license and bonding, if necessary in your area. Contact your local chamber of commerce and local city and state officials in order to find out what is required.
- Tools and parts suppliers you can trust. That also means building relationships with manufacturers because some parts may not be available outside of the manufacturer.
- People skills. You'll be entering peoples' homes and sometimes taking their sets away.
- Quick assessment skills to troubleshoot and fix sets on the spot or being able to determine if a set is salvageable before having to transport it to your workshop
- Advertising in print and online to promote your business. A website can also help promote your home business, as most potential customers may search online for TV repair services in their area.

TV Repair: Real-Life Example

Briargrove TV Service, a two-person operation owned by Jerry Mantaring in Houston, earns an estimated $50,000 annually, according to a Manta.com business profile.

ADVICE ON JOB HUNTING AND YOUR CAREER

Do Your Homework

There's no excuse for ignorance about the television industry when information can easily be found online or is just a phone call away. Let your fingers do the clicking, saddle up to the Internet and research what you want to know about a particular station, market or job. Key words such as "careers in television" entered into a search engine would be one basic route. Go to station Web sites and tap into the staff bios. Then go to the "Jobs Available" section. You'll see how others progressed and how you might tailor your own career path. This research will also help you create a cover letter that presents you as informed and savvy about the station's history and "personality," which allow you to drop "buzz words" that might open that front door.

Cover Letters

Don't just address a cover letter to a generic "Dear Sir or Madam." Call and find out who at the station is hiring and personally address the letter. Make sure you have the correct spelling – you don't want to be an applicant too lazy too have made an effort to ask the right questions. After all, this is a journalistic business and it pays – as it does anywhere – to get it right the first time. A well-written cover letter won't necessarily get you in the door, but a poorly crafted one will keep you out, especially if you're after jobs in areas such as news, programming, creative services, public relations, accounting and sales. Cover letters must accompany a resume and they should reflect your knowledge of the station, the job you seek, and present you in a way that stands out from the next hungry competitor.

Resumes

Many station managers keep certain resumes and cover letters on hand as examples of what not to do. Some favorite faux paus? Cover letters and resumes that lack such basic contact information as a phone number; ones addressed "To Whom It May Concern;" and those loaded with spelling and grammatical mistakes.

Then there are the content-weak letters that breathlessly note, "I just want a job in television, any job – what do you have to offer?;" the ones where every paragraph starts with an "I...;" and those ego-centric ones that read "P.S. – you can reach me at (111) 234-5678," and actually believe they'll get a call back. If you're not great at constructing a cover letter or resume, get help from a parent, teacher, friend, colleague or professional. Know the importance of follow up.

The Interview

Once you secure an interview, there are a few more rules you will want to follow, including following up to confirm the interview the day before. Arriving not only on-time, but early-,is a good idea. Dress for success, and be sure that your attire is appropriate for a professional office. Double-check in the mirror for spots on the tie, lost buttons, dangling threads and don't forget to cut the tags off that new suit! Remember, this is television. Appearances count.

ADVANCING IN THE INDUSTRY

Examine the resumes of reporters and anchors and you'll discover they aren't nicknamed "glitter gypsies" for nothing. Non-accidental tourists, they go where the bigger and better jobs are, especially younger talent, who often move every year or so to achieve their goals.

Some career moves may puzzle those unfamiliar in the industry. Why, for example, would someone take a job as part-time weekend reporter when he or she is already an anchor? While this may seem like a demotion, if you compare the stations' markets you realize that the "anchor" position was in the nation's 47th market, while the weekend anchor position is at a top-20 station. Thus, because there are more opportunities for advancement at the second station it is actually a smart career choice.

When reporters do reach a market where they're comfortable with the size, lifestyle and pay, they may well put down roots even if anchor jobs appear out of reach. Why? This is still a business where appearances count and stability eventually wins out over ego. Those seeking management positions also need to move in order to move up. That may mean seeking a post with a different station group or actively campaigning for promotion within the same corporate structure – which still might mean transferring to a new part of the country. Be versatile. Be open. Be willing to take on new tasks without ignoring the tasks in front of you. Be curious. Best of all, be the best you can be to move up.

A WORD ABOUT SALARIES

Poultry king Frank Perdue gave a great piece of advice to graduating students seeking to enter the work world and it applies to every job category: "Don't go for the job that offers the most pay. Go for the job that offers the most advancement." That's difficult advice when buddies are out making $40,000 and up a year. Will they tap out at $40,000, too? Are they in a career that they love?

A WORD ABOUT STATION OWNERSHIP

In terms of employment, station ownership may not affect entry-level position job seekers. However, as with any company, a station's resources (financial, technological, etc.) and structure will have an impact on the various positions at the television station – from what positions are available to the job responsibilities of each position. Thus, it is important to know and understand the differences in station ownership.

Most commercial television stations are owned by large media conglomerates. Smaller, family-held stations are the exception rather than the rule today as mergers and acquisitions have brought many stations under one large corporate umbrella.

There are several types of commercial stations. These include stations that are owned and operated by the networks (O&Os), stations that are network affiliates of a network (ABC, CBS, Fox, NBC, Univision, PAX TV,

WB, etc.) and independent stations that are not affiliated with any network. Large media groups often own affiliates of several different networks.

Network-owned and operated stations carry network programs, have a good deal of local autonomy and ultimately report to their network's station group management. Network affiliates also enjoy local autonomy and have a contractual relationship with a network for the network to supply them with programming. Both affiliates and O&Os carry syndicated programming supplied by vendors such as King World and Twentieth Television, among others, which are either independently owned or exist as companies within the same corporate family. Independent stations carry syndicated programming and/or locally originated fare.

Public television stations are often owned or managed by local governments, universities and/or school districts. The majority of these stations are members of PBS receiving a large portion of their programming from this service. The remainder of their programming day is filled with syndicated and locally originated programming. Public stations cannot air advertising and depend on the funding they receive through viewer support, corporate underwriting and government funding.

Whether the entry-level position you find is with a small station or a large conglomerate, your job performance, in large part, will ultimately determine your upward mobility. It is also important when considering a career in television to know that the industry is mobile and relocation is often a reality.

THE GENERAL ADMINISTRATION DEPARTMENT IN TELEVISION

General Manager

A station's general manager (GM), who often holds the additional title of president or vice president in larger station groups (and may also be the owner in a small cluster), is the chief operating officer at the station. All departments typically report directly to the general manager. The general manager sets the work ethic, tone and pace of the station. TV general managers predict market trends; set budgets and performance expectations; ensure the station is in compliance with all equal employment opportunity (EEO) and Federal Communications Commission (FCC) requirements; and have overall responsibility for maintaining a station's reputation and market image. TV general managers have final say on station hires and keep both on-air and behind-the-scenes talent in step with overall station directives. They create a clear vision for the station and the blueprints for success that every staffer is expected to follow.

The bottom line stops with the GM, who is charged with growing the station's profitability. Advertising revenue is crucial to a television station's financial security and, as a result, relationship-building with advertisers and potential advertisers is a significant area of the GM's responsibilities. The GM is heavily involved in the station's sales efforts and is generally expected to participate in sales presentations along with executive members of the sales staff. GMs find creative ways to profit and succeed despite obstacles.

Every television station cultivates its own public image, based on its local market and community needs. As a result, general managers are committed to involving the station in community events and public affairs. Television stations are visible leaders in their markets, and a many choose to deliver on-air editorials to voice opinions about community issues at large.

General managers are typically recruited from within the advertising sales department, although news directors – especially those with experience in other aspects of the business – are also potential candidates for this position. The average length of stay in a market for a GM is five years. A traditional advancement path for a general manager is to leave a smaller market station for a larger market, either within his/her own company or by competitor recruitment (outside the company). General managers are often recruited from their stations by corporate offices for positions at company headquarters.

Station Manager

The position of station manager was originally designed to alleviate some of the pressures placed on the GM. Many stations are eliminating, or intend to eliminate, this position, thus the number of job opportunities for this title are diminishing. Station duopolies (stations that operate in markets where two television stations are owned by a single owner) may be an exception, where given the increased demands on the GM, the position of station manager remains necessary. In these instances, the GM focuses on long-term strategies, budgets, community networking, business building, advertiser relations and policy making, while the station manager concentrates on the day-to-day operations, interfacing with department heads and handling personnel issues.

Human Resources

In those stations where there is a station manager, the station manager often oversees the day-to-day human resource (HR) functions. Otherwise, the HR functions will fall to the general manager, who at larger stations will have a human resource staffer. Years ago, stations often had a senior human resources professional in this role, but this senior position is now often found at the corporate level, leaving more junior people at the individual stations within a group.

Executive Assistant

Executive assistants handle executive/senior-level correspondence and phones; field viewer questions and complaints; manage the executive appointments calendar; process billing; handle the paperwork related to EEO and FCC reporting and, in general, try to make the day go more smoothly for the GM. Don't confuse the abundance of clerical duties with a lack of importance in this position. The executive assistant is often the person who determines who gains access to the GM, and thus this administrative position can be a powerful role at the station.

Business Manager and Controller

Some stations have separate positions for business manager and controller while other stations may combine these two positions since the duties for each position are similar: maintaining the station's financial well-being. These financial experts report directly to the general manager and at larger stations manage a business staff that oversees a station's accounting. They also make decisions in concert with the station's department heads and the general manager about large equipment purchases and expenditures.

Most business managers are Certified Public Accountants and many hold MBA degrees. All have high degrees of financial acumen in accounting and financial management as well as the necessary computer and data processing skills.

THE PUBLIC RELATIONS AND COMMUNITY AFFAIRS DEPARTMENT

Public Relations Director

A public relations (PR) director for a television station functions the same as any other PR person, the difference being that the product he or she promotes is the television station. Skilled in written and oral communications, the public relations director manages public perception of the station, interacts with the media and serves as an official spokesperson. Special station events are often arranged by this department to further connect with the community or advertisers and could be the domain of the public relations director or the community relations manager (who may hold the title of public service director).

The job of public relations director could be a permanent staff position or handled on a contractual basis by an outside firm. At stations lacking a public relations director, the duties are often added to the job of community relations manager or public service director. In certain cases, the general manager handles media queries and interfaces with the press.

Public Service Director or Community Relations Manager

In addition to the public relations director, the station's community liaison is the public service director or community relations manager, who works to involve the station in the community through sponsorships of charity events or other visible causes, including those that could provide advertiser tie-ins. Public service directors/community relations managers receive countless public service announcements (PSA) submitted by groups hoping to get their non-commercial messages on the air and their causes promoted "for free." A number of public service directors/community relations mangers write, produce and even edit local public affairs programs and may even serve as an on-air hosts. Most people employed in these positions have college degrees and majors in communications, English, Liberal Arts or Political Science. Public relations directors have specific experience in the public relations field and may have worked before at an advertising and public relations agency, held a PR post within a corporation or come from print or broadcast journalism.

Public service directors/community relations managers are experienced in broadcast production and may or may not have on-air skills. They might have transferred from radio, worked at a post-production house or served as an on-air talent in TV or radio and somewhere along the way acquired experience in writing and producing for television.

Speakers Bureau Manager

A part-time position, if it exists at all, the speakers bureau manager deals with public requests for talent appearances at various events. What does and does not get scheduled is a function of individual talent preferences and what causes the station has adopted as its own. This person may or may not have a college degree and could have been promoted up from a job as department assistant or intern.

THE PROGRAMMING DEPARTMENT

Programming Director

The programming director or programming manager makes certain programming information is fed to traffic, sales and other pertinent departments. Any changes are communicated both in house and externally to the TV listings services and local newspapers. Maintaining programming equilibrium versus negotiating for new product is part of the daily responsibilities. Stations in larger markets, especially those that are part of a large media conglomerate, have downgraded or eliminated this position because many programming decisions are now made at the corporate level, on behalf of multiple stations.

THE ENGINEERING DEPARTMENT

Director of Broadcast Operations and Engineering (BO&E)

The staff of the Engineering department keeps the television station on the air and the station's physical property in good shape. Rapidly evolving technology has compressed the workforce in this area, making multi-tasking more essential and some jobs obsolete.

Head of this department is the director of broadcast operations and engineering (BO&E), who could also be called chief engineer. This is a hybrid position that's half technical, half personnel management, overseeing -- in the words of one BO&E director – "everything from the toilet paper to the tower."

BO&E heads figure out what new equipment the station needs to keep competitive and work the angles to get the best deals. While a number of news departments are still video-tape based, most are converting to non-linear, server-based technologies. Digital technology is replacing analog, but stations still need to deal with video's legacy: all the information still physically stored in this format versus compressed computer files.

General Broadcast Operations and Engineering Positions

Although titles may shift station to station, the following positions also comprise the BO&E team. The engineering manager supervises such areas as equipment maintenance, installation, repair of anything from the tower to news vans, and the technicians who accomplish these tasks. Physical maintenance of the property – its landscaping and grounds, janitorial needs, mailroom, security and front desk reception is the domain of the building supervisor who manages the staff working in these areas.

A broadcast operations manager oversees the scheduling of studio time and some of studio personnel associated with the physical side of on-air performance such as show directors; camera operators (most studio cameras are now robotic); tape room operators who manipulate video tape; master control supervisors or chiefs who oversee the master control operators; the audio board operators and switchers; teleprompter operators and more.

Those working in master control are very familiar with FCC standards and practices related to video levels and take action to ensure the integrity of the station's on-air technical performance. It is here that the station's daily log is programmed into the spot playback system, tapes are loaded in the proper sequence, air variances are corrected prior to airing of any commercials or other inputted programming. Master control operators log transmitter readings, align satellite receiver equipment and record programming feeds as logged on the satellite as well. It's the news technical operations manager who schedules shifts for the photographers or videographers and other people involved in the tangible editing of the news product, whether they're working with video tape, film, high-definition or other digitally formatted mediums. Crew chiefs in the photography and editing areas report to the news tech ops manager, or technical resources manager, and in turn manage down the line. Engineering technicians of all kinds are needed to keep equipment humming, from the phone system and satellite network, to reporter laptops and the plumbing.

Entry Level Engineering Positions

Entry-level jobs in the Broadcast Operations and Engineering department include trainees for such positions as teleprompter operator, master control operator or video editor. A high school education is required with some college preferred although trade school and/or prior field experience count. It depends on who's hiring and market size.

THE TRAFFIC DEPARTMENT

Operations Manager

Operations managers head the traffic department, whose responsibility is to set the advertising schedule that is crucial to maximizing a station's revenue and keeping commercials on the air. Logical thinking and good task juggling abilities are critical in this position. An operations manager's job is highly stressful; his or her performance could mean the difference between the sales department achieving its budget or not. Operations managers ensure the traffic department keeps an accurate station log – the master listing of all scheduled programs and commercials expected to air throughout each broadcast day.

One of the television station's major goals is to ensure every spot (i.e. television commercial) is sold for the highest amount while the perishable inventory – or available time to run commercials – is maximized. An operations manager knows how to manipulate the schedule to meet both the client's and station's needs.

For example, consider a situation where a movie company wants to buy spots promoting a film opening in just a few days and is willing to spend top-dollar, but the television station has sold out its inventory for that weekend. The operations manager would identify clients that may be willing to alter their advertising schedule, sometimes offering additional incentives such as extra spots added elsewhere at no charge, or offering up other programs with air time left to sell in hopes of clearing out that inventory while at the same time benefiting the client.

Other Traffic Department Positions

Larger stations may structure the traffic department as follows: at the head of the department is the operations manager who reports to the DOS and directly supervises the traffic manager. The traffic manager oversees the traffic supervisor who is in charge of the other members of the traffic team: the traffic assistants (also known as order entry coordinators or log editors or copy coordinators).

Traffic assistants create and manage the next day's commercial logs, make sure copy is properly entered; type quickly and accurately, and perform other clerical duties to keep sales running.

THE RESEARCH DEPARTMENT

Market Research Director

A market research director interprets a station's ratings and supporting research to cast the station most favorably against its competitors, and assists in the creation of sales materials. Sometimes he or she participates in sales presentations to existing and potential clients.

A station's market research director may supervise a staff or the position may not exist at all, with the corporate office providing the local stations the necessary ratings and research data. The credentials and number of years in the field vary as well for this position. Often, larger station groups that maintain corporate research departments distribute information as necessary to stations within the group. Thus, the more heavily credentialed people are at the corporate level, eliminating the need for duplication on the local levels. Not only does this type of job streamlining occur in the area of market research, but it also happens in human resources and programming areas.

THE NEWS DEPARTMENT

News Director

A news director sets newsroom journalistic standards that govern who, what, where, when and why a particular news event is covered – or not covered. News directors make certain the journalistic integrity of the station remains uncompromised, map out coverage strategies and plan assignments that showcase the talent in ways compatible with the station's image goals. This is a coveted position to which many ambitious journalists aspire (unless they want to go even higher and become the GM).

Typically long-range thinkers, news directors identify future scheduled events and plan coverage around them that allows the station to take ownership of the event and shine brighter than the competition. The event could be an elite sporting event, political debate, huge charity spectacular or other special function. News directors must also plan for the unexpected and are expected to react well in a crisis, keeping the news staff focused and committed to performing under extreme tension and stress.

Management duties for this position include financial aspects such as developing budgets that allow leeway for overtime when the unexpected occurs.

News directors understand that while their foremost focus is on journalistic excellence, television is also a business of revenue and ratings. Staff within the sales department will often ask news directors to cover client-generated topics or events, and it is up to the news director, sometimes in consultation with the general manager, to determine potential conflicts of interest in these situations.

Assistant News Director

The assistant news director is often next in line to become the station's news director and has a solid news background and management skills. The assistant news director is responsible for implementing the news director's vision and questioning the content, accuracy and fairness of stories to maintain balance and station credibility.

Managing Editor

The managing editor – if a station has this position – manages the more day-to-day operations of the newsroom.

Business Manager for News/News Operations Manager

The business manager for news or the news operations manager takes care of the financial details for the News department, tracking budget information, overtime pay, departmental supplies and more, keeping in close contact with the news directors and assistant news directors.

Executive Producer

Executive producers (EP) have overall responsibility for their individual news program or show. Executive producers direct shots, approve show content and make sure their program or show is on track. Those in this post think in 30-second intervals and are prepared to immediately switch gears in a crisis, such as a lost feed, no-show guest or any other unforeseeable event. Executive producers must be creative thinkers and able to create a newscast with a defined personality that stands out and connects with viewers.

Senior Producer

Each news program or show in the larger markets has a senior producer who creatively shares with the executive producer the responsibility for determining the composition and flow of the newscast. Senior producers work closely with the news management, anchors, reporters, editors and the technical crews; they research news story lines, write, select video and set interviews for news stories.

Producer and Assistant Producer

Producers help write for the shows and perform the detail-oriented tasks that, when properly done, can mean the difference between a great show and a mediocre one, or convince a source to talk on air. Assistant or associate producers work to make the lives of producers easier by handling any and every task assigned – whether it is confirming a guest, driving across town for an essential prop or doing background research.

News Writer

News writers do just this: write the news. Not many stations have more than one dedicated full-time to this task; often producers do double duty as show writers.

Futures or Planning Editor

Futures or planning editors forecast the station's upcoming coverage by making decisions about what the station might cover in the next days or weeks. Advance planning allows the station to develop a story rather than just reacting to a news event.

Assignment Desk Chief

A seasoned veteran, the assignment desk chief works with others in news management to plan well-rounded news coverage that fits with the "personality" of the station. Assignment desk chiefs maintain extensive working files of possible story ideas, including crucial background information, and have superb contacts. They also oversee the scheduling and assignment of news coverage, order satellite time as needed, and take active roles in planning in-depth series and special coverage. Assignment desk chiefs may also function as writers and producers as needed.

News Assignment Desk

The news assignment desk is one of the most important posts in the newsroom. In smaller stations, the staff in this area may be entry-level (typically the smaller the station, the younger the faces in this department).Staffers here could work days, nights or weekends. They man the news tip phone lines; listen to police dispatches; read newswires, emails and press releases; stay in contact with key sources; and assist with assigning stories, scheduling reporters and photographers. They often are the "first in the know," and must exercise considerable news judgment to separate fact from fiction.

Special Projects Editor

Special projects editors work on long-term assignments that are more in-depth and may evolve into a series of reports. They work closely with the news talent, and their work is often highlighted during the crucial sweeps, during the key ratings months.

News Assistant

News assistants serve as the right-hands of the news director, fielding phone calls, interfacing with other departments to get essential intradepartmental information and more.

News Anchor/On-Air Talent

News anchors are the foundation upon which the news team is built –

they are the most recognizable station ambassadors to the public. Generally news anchors have worked their way up through the reporting ranks, although some may have started on the production side. Competent journalists, news anchors write and package their own special reports as well as read copy prepared by other team members. With today's technology, more anchors are stepping out from behind the anchor desks to do "live remotes" and report on the higher profile stories. At smaller stations, weekday news anchors have formal reporting duties incorporated into their roles. At larger stations, it's generally only the weekend anchors that have this double duty.

Sports Director

Sports directors report on local and national sporting events. Due to increased competition from the cable sports' outlets, sports directors often now have less air-time than when only the "big three networks" dominated the airways. Better, more concise reporting, and an ability to bond with the sports teams is crucial to how viewers now relate to a station's sports director.

Weathercaster/Meteorologist

Weathercasters/meteorologists report and forecast the local area's weather conditions. Newer entrants into this area of news generally hold degrees in meteorology. Their on-paper credentials become a marketable asset used in station promotions. Yet degrees don't matter if you can't deliver the weather in a way that's credible, informative and resonates well with viewers. With Doppler radar installed on-site at many stations and computers improving the content, accuracy and scope of weather-related information, success in this position is a function of how well the weathercaster/meteorologist can master the weather equipment and integrate that information into his/her reports.

THE SALES AND MARKETING DEPARTMENT

Director of Sales

The term general sales manager (GSM) is sometimes substituted for director of sales (DOS). In large markets, the DOS may supervise a team of GSMs and/or hold a dual title for one of the stations within the group. Along with the news director and business manager, the director of sales is part of the general manager's "inner circle." This is an extremely demanding position and requires exemplary management skills.

A DOS has daily contact with the general manager, and works with other sales managers to ensure the station is maximizing all available traditional revenue opportunities in the market – from both local and national ad agencies – while developing new business opportunities from both. Directors of sales price and manage the perishable inventory; do budget forecasts; issue sales projections; coordinate projects with other department heads; motivate the team; and inspire everyone to give 150 percent.

General Sales Manager

At some stations, the general sales manager (GSM) is a distinctly different position than director of sales. GSMs have complete knowledge of sales forecasting and planning and are strong motivators. Along with the DOS, they are long-range thinkers, have excellent presentation and interpersonal skills and excel at client networking. GSMs develop monthly and quarterly revenue-generating proposals, establish individual sales goals and hire and manage the sales teams.

National Sales Manager

National sales managers (NSM) manage a television station's sales accounts. A national sales manager "works the phones," deals with the sales representatives at national advertising firms and travels to the advertising epicenters including New York, Chicago, Los Angeles, Atlanta and Dallas. National business is very fast-paced. Sales representatives at advertising firms working with NSMs handle a host of stations and must make decisions quickly, thus it is imperative that national sales managers have all the station's updated information, are on top of all pending business and are accessible at all times to negotiate buys. Stations lacking a national sales manager give the responsibilities of this position to the GSM.

Marketing Director or Director of Non-Traditional Revenue

The position of marketing director or director of non-traditional revenue is a more recently created position within television stations, added as more stations seek non-traditional revenue sources. The performance bars for these positions continue to rise. Marketing directors continually develop and implement creative, out-of-the-box ways to generate revenue. One successful example of a non-traditional revenue source is staging an "Ask the Pharmacist" day during which pharmacists associated with a professional network come into the station and dispense advice to callers. This group discussion is not broadcasted, but promotional spots encouraging viewers to call the station are, and the pharmaceutical client pays for the promotional costs. The marketing director may also work with a local charity, sponsoring an event with commercial TV tie-ins and/or printed collateral materials. Depending on the station, marketing directors may have a support staff and/or one or two account executives assigned to help in their area, and also may work with the production people in creative services to produce commercials for new advertisers.

Local Sales Manager

Local sales managers (LSM) closely supervise the frontline sales staff consisting of account executives or account managers – whatever term that station prefers. If a station does not have a general sales manager, the local sales managers acquire the duties of this position.

Account Executive

Account executives (AE), who also may be called account managers, comprise the local sales team that generate advertising revenue by calling on established agencies, cold-calling new prospects and convincing potential clients of the merits of television advertisements. Account executives often have irregular hours and work evenings and weekends meeting with clients.

An innate sales ability and a positive attitude are essential in this position. Most account executives receive a commission based on7 their sales volumes, and thus earnings will fluctuate depending on sales performance. This isn't necessarily a job for those persons who want a consistent monthly paycheck. But for successful account executives, the potentially outstanding financial rewards make this an attractive position.

Sales Assistant

Sales assistants keep the communications flowing within the sales department, working with managers and account executives to create sales presentations and other sales support materials. They are adept at a number of computer programs (e.g., PowerPoint), input sales orders, troubleshoot client and agency phone calls, help track and gather articles that support the sales cause and may even plan the details of sales entertainment events.

THE CREATIVE SERVICES DEPARTMENT

CREATIVE SERVICES OR PROMOTIONS DIRECTOR

Creative services or promotions directors develop and oversee the station's image marketing, direct all advertising, promotional and contest initiatives of the station, as well as daily promotion of the news product. They work to properly brand the station in concert with the general manager, who sets the overall station image goals and objectives.

Imaginative and visionary, creative services directors have strong artistic backgrounds and are well-versed in copywriting and editing as well as audio, visual and print production. Supervision of the art department is part of their role, along with directing the activities of an outside ad agency for those larger stations with that relationship.

Many Creative Service Directors do the media planning and buying and all are responsible for staff hiring, budget projections and management in their supervisory area. In smaller markets, especially, they may do hands-on editing and should have a degree of knowledge of how the editing process works.

Promotions Manager or Audience Promotions Manager

Promotions managers or audience promotions managers write, create and coordinate with other Creative Services staffers, the on-air promotions for station-sponsored special events. They work with sales to create client, revenue-driven promotions and also work with news to create daily topicals. Their department (which could consist of just one person) also maintains the station's banners and other promotional items and giveaways used at station events. Stations with bigger staffs have writers/producers assigned to work under the promotions manager, often dividing writing and promotions into two separate focus areas to maintain the separation of news and sales.

Writer/Producer

Writers/producers (one of which may double as Web master) – assigned to news promotion write and produce topicals and spots that highlight the news and programming products. They are also involved in creating a station's branding campaign. While familiar with shooting techniques, writers/ producers generally do not operate this equipment themselves. Writer/ producers serving the sales departments write commercials for those clients who don't have an outside ad agency, and produce and shoot these spots themselves. Writers/producers have excellent creative writing and presentation skills as well as great talent for the visual elements. They have experience in such graphics programs as Avid, Adobe Photoshop, Adobe Premier and Final Cut, and editing with non-linear and linear platforms.

Post-Production Editor

Post-production editors take rough cuts (raw footage) and polish them into final, on-air promotional materials. This entails adding music, graphics and other sound and sight elements as necessary to make the finished product even more compelling. Post-production editors may work on a station commercial, a news package or a public affairs program.

Station Announcer

Station announcers generally are hired on a contractual basis to do voiceovers for news promotions.

Art Department Positions

An art director (AD) oversees the graphics area and has specific knowledge of what works for TV and what doesn't. Art directors work with news to give more visual impact to the news product by using graphics or other special effects such as animation and deal with outside vendors on such projects as set design, signage and station image branding materials including print ads and collateral. Art directors also supervise electronic graphic artists working for the station.

A Word About the Creative Services Department

Permanent jobs often missing from this department are the make-up person and the props person. These tasks, when needed, are usually filled by independent contractors (i.e. freelancers). People hired in the Creative Services department usually have Bachelor of Arts degrees with majors in these most common areas: English, Communications, Liberal Arts, Political Science, History, or, as in one known case, anthropology. All possess excellent English and grammar skills.

Most schools did not offer communications majors until the mid-1970s. Those who graduated after this period have the added advantages of specific training in focused areas of the communications field, from concentrations in broadcast journalism to electronic graphic arts and television editing and production. Many colleges require specific knowledge of such editing programs as Final Cut. Familiarity with this program is a real plus in this business, for certain employees in the creative services area who lack these skills are required to get up to speed and learn how to edit. Art department staffers have specific training in electronic graphic arts; just a freehand background doesn't make the grade. The typical path to advancement in the Creative Service department is through the ranks; most have experience at more than one television station if they are employed at a station in the larger markets.

2

Modern Cable Television Technology

Cable television is one of the most important technologies available to the average citizen. While television has become the preferred source of information for the masses, cable television has been the reason viewers have many choices. In fact, if cable had to be described with just one word, that word would have to be:

HOW CABLE TELEVISION WORKS

In the 1950s, there were four television networks in the United States. Because of the frequencies allotted to television, the signals could only be received in a "line of sight" from the transmitting antenna. People living in remote areas, especially remote mountainous areas, couldn't see the programs that were already becoming an important part of U.S. culture.

In 1948, people living in remote valleys in Pennsylvania solved their reception problems by putting antennas on hills and running cables to their houses. These days, the same technology once used by remote hamlets and select cities allows viewers all over the country to access a wide variety of programs and channels that meet their individual needs and desires. By the early 1990s, cable television had re ached nearly half the homes in the United States.

Today, U.S. cable systems deliver hundreds of channels to some 60 million homes, while also providing a growing number of people with high-speed Internet access. Some cable systems even let you make telephone calls and receive new programming technologies! In this article, we'll show you how cable television brings you so much information and such a wide range of programs, from educational to inspirational to just plain odd.

The earliest cable systems were, in effect, strategically placed antennas with very long cables connecting them to subscribers' television sets. Because the signal from the antenna became weaker as it traveled through the length of cable, cable providers had to insert amplifiers at regular intervals to boost the strength of the signal and make it acceptable for viewing. According to Bill Wall, technical director for subscriber networks at Scientific-Atlanta, a leading maker of equipment for cable television systems, limitations in these amplifiers were a significant issue for cable system designers in the next three decades. "In a cable system, the signal might have gone through 30 or 40 amplifiers before reaching your house, one every 1,000 feet or so," Wall says. "With each amplifier, you would get noise and distortion. Plus, if one of the amplifiers failed, you lost the picture. Cable got a reputation for not having the best quality picture and for not being reliable." In the late 1970s, cable television would find a solution to the amplifier problem. By then, they had also developed technology that allowed them to add more programming to cable service.

Adding Channels

In the early 1950s, cable systems began experimenting with ways to use microwave transmitting and receiving towers to capture the signals from distant stations. In some cases, this made television available to people who lived outside the range of standard broadcasts. In other cases, especially in the northeastern United States, it meant that cable customers might have access to several broadcast stations of the same network. For the first time, cable was used to enrich television viewing, not just make ordinary viewing possible. This started a trend that would begin to flower fully in the 1970's.

The addition of CATV (community antenna television) stations and the spread of cable systems ultimately led manufacturers to add a switchto most new television sets. People could set their televisions to tune to channels based on the Federal Communications Commission (FCC) frequency allocation plan, or they could set them for the plan used by most cable systems. The two plans differed in important ways.

In both tuning systems, each television station was given a 6-megahertz (MHz) slice of the radio spectrum. The FCC had originally devoted parts of the very high frequency (VHF) spectrum to 12 television channels. The channels weren't put into a single block of frequencies, but were instead broken into two groups to avoid interfering with existing radio services.

Later, when the growing popularity of television necessitated additional channels, the FCC allocated frequencies in the ultra-high frequency (UHF) portion of the spectrum. They established channels 14 to 69 using a block of frequencies between 470 MHz and 812 MHz.

Because they used cable instead of antennas, cable television systems didn't have to worry about existing services. Engineers could use the so-called mid-band, those frequencies passed over by broadcast TV due to other signals, for channels 14-22. Channels 1 through 6 are at lower frequencies and the rest are higher. The "CATV/Antenna" switch tells the television's tuner whether to tune around the mid-band or to tune straight through it.

While we're on the subject of tuning, it's worth considering why CATV systems don't use the same frequencies for stations broadcasting on channels 1 to 6 that those stations use to broadcast over the airwaves. Cable equipment is designed to shield the signals carried on the cable from outside interference, and televisions are designed to accept signals only from the point of connection to the cable or antenna; butinterference can still enter the system, especially at connectors. When the interference comes from the same channel that's carried on the cable, there is a problem because of the difference in broadcast speed between the two signals.

Radio signals travel through the air at a speed very close to the speed of light. In a coaxial cable like the one that brings CATV signals to your house, radio signals travel at about two-thirds the speed of light. When the broadcast and cable signals get to the television tuner a fraction of a second apart, you see a double image called "ghosting."

In 1972, a cable system in Wilkes-Barre, PA, began offering the first "pay-per-view" channel. The customers would pay to watch individual movies or sporting events. They called the new service Home Box Office, or HBO. It continued as a regional service until 1975, when HBO began transmitting a signal to a satellite in geosynchronous orbit and then down to cable systems in Florida and Mississippi. Scientific-Atlanta's Bill Wall says that these early satellites could receive and retransmit up to 24 channels. The cable systems receiving the signals used dish antennas 10 meters in diameter, with a separate dish for each channel! With the beginning of satellite program delivery to cable systems, the basic architecture of the modern cable system was in place.

As the number of program options grew, the bandwidth of cable systems also increased. Early systems operated at 200 MHz, allowing 33 channels. As technology progressed, the bandwidth increased to 300, 400, 500 and now 550 MHz, with the number of channels increasing to 91. Two additional advances in technology — fiber optics and analog-to-digital conversion — improved features and broadcast quality while continuing to increase the number of channels available.

The Glass Cable

In 1976, a new sort of cable system debuted. This system used fiber-optic cable for the trunk cables that carry signals from the CATV head-end to neighborhoods. The head-end is where the cable system receives programming from various sources, assigns the programming to channels and retransmits it onto cables. By the late 1970s, fiber optics had progressed considerably and so were a cost-effective means of carrying CATV signals over long distances. The great advantage of fiber-optic cable is that it doesn't suffer the same signal losses as coaxial cable, which eliminated the need for so many amplifiers. In the early fiber-optic cable systems, the number of amplifiers between head-end and customer was reduced from 30 or 40 down to around six. In systems implemented since 1988, the number of amplifiers has been further reduced, to the point that only one or two amplifiers are required for most customers. Decreasing the number of amplifiers made dramatic improvements in signal quality and system reliability. Another benefit that came from the move to fiber-optic cable was greater customization. Since a single fiber-optic cable might serve 500 households, it became possible to target individual neighborhoods for messages and services. In the 1990s, cable providers found this same neighborhood grouping to be ideal for creating a local-area network and providing Internet access through cable modems.

In 1989, General Instruments demonstrated that it was possible to convert an analog cable signal to digital and transmit it in a standard 6-MHz television channel. Using MPEG compression, CATV systems installed today can transmit up to 10 channels of video in the 6-MHz bandwidth of a single analog channel. When combined with a 550-MHz overall bandwidth, this allows the possibility of nearly 1,000 channels of video on a system. In addition, digital technology allows for error correction to ensure the quality of the received signal. The move to digital technology also changed the quality of one of cable television's most visible features: the scrambled channel.

The first system to "scramble" a channel on a cable system was demonstrated in 1971. In the firstscrambling system, one of the signals used to synchronize the television picture was removed when the signal was transmitted, then reinserted by a small device at the customer's home. Later scrambling systems inserted a signal slightly offset from the channel's frequency to interfere with the picture, then filtered the interfering signal out of the mix at the customer's television. In both cases, the scrambled channel could generally be seen as a jagged, jumbled set of video images.

In a digital system, the signal isn't scrambled, but encrypted. The encrypted signal must be decoded with the proper key. Without the key, the digital-to-analog converter can't turn the stream of bits into anything usable by the television's tuner. When a "non-signal" is received, the cable system substitutes an advertisement or the familiar blue screen.

THE CURRENT PICTURE OF CABLE IN THE UNITED STATES

Choice is made possible by cost effective massive bandwidth for delivering programming to almost seventy percent of the 108.4 million households having television receivers in the United States. Since cable passes over ninety percent of television households, nearly everyone has the opportunity to subscribe to cable. In the United States, more people have television receivers than telephones, and most of those people have cable service.

Cable television is not the only multi-channel video provider. Direct Broadcast Satellite, DBS, has become a major alternative with about 22 million subscribers. This is about the same number of subscribers as served by the largest cable Multiple System Operator, MSO, Comcast and about double the second largest MSO, Time Warner Cable. Cable penetration of television households, TVHHs, has declined slightly over the last few years from a peak of 69.2% in 2001 to 67.7% in 2003, losing some market share to DBS. At the same time, cable's share of the public's viewing has grown while the Broadcast network's share has steadily fallen. In 2003, basic cable had 57% of TVHHs while broadcast had 38%. 1996 was the year the Broadcast network share dropped below 50%.

There are over 372 "channels" on the satellites that feed nearly ten thousand cable headends. An additional 127 planned "channels" are in the start up mode. Cable systems must select some of the programs they will make available to their subscribers from this set since no analog cable system has sufficient capacity to carry them all and many digital systems would be hard pressed to maintain an attractive analog offering while taking all of these channels and still providing other services. In addition, cable systems carry the local broadcasts and sometimes programs locally originated or designated for local citizen access. In some cable systems, the local schools, library, or local government have channels to deliver programs of special interest. In this way, the needs of a democratic society are served while citizens become well-informed, and entertained as well.

The broadband cable infrastructure can deliver an appealing analog service while simultaneously adding hundreds of digital channels. This hybrid service is important for the transition to digital television since an abrupt change over from analog to digital television would be chaotic. Most citizens will continue to enjoy analog television for many years to come. Even those who purchase a digital television receiver for their principal viewing room will have analog receivers in the rest of the house. None the less, there is increasing interest in converting to all digital video or at least mostly all digital video. High Definition Television, HDTV, is generating a lot of interest (and a lot of confusion) among consumers. HDTV is available on cable to over 84% of cable subscribers. More than 16 cable networks are producing HDTV programming. Sports seem to be the main driver.

The cable infrastructure is inherently two-way, requiring only modest additional investment and maintenance to implement. When activated, this capability facilitates interactive cable services and other non-traditional cable services such as high speed data services, telephony, Internet telephony (termed Voice of Internet Protocol, VoIP) and even video telephony.

High speed data services using cable modems are currently available to nearly 90% of cable subscribers. There are more than 16 million cable modem subscribers. Digital video cable subscribers number over 22 million. The cable industry spends over ten billion dollars a year on infrastructure, having peaked at over 16 billion in 2001. Cable systems spend over 11 billion a year purchasing programming. The average monthly price of expanded basic service is $36.59. Expanded basic is the most popular level of service and is one level up from basic. Basic includes the mandatory broadcast and franchise required local channels and occasionally a few others. Very few subscribers take just basic. The cable industry has revenues from subscribers of $52 billion and advertising revenues of $16 billion.

Current areas of high interest encompass the "triple play" offering of video, voice, and data. Digital video includes Video on Demand, VOD, HDTV, and Digital Video Recorders, DVRs (also called Personal Video Recorders, PVRs). DVRs employ a computer hard disk to record tens of hours of programming while providing the Video Cassette Recorder, VCR, features of pause, fast forward, and rewind. An appealing feature for subscribers and an appalling capability for advertisers is found in the DVR's ability to skip commercial messages. The cable industry is struggling with how to serve subscriber's desires for this capability without ruining the important advertising revenue stream.

The success of cable has stimulated other broadband competitors, all using the programming services made possible by cable's investments and commitments to provide audience. The cable industry has financed and developed these program services, tested their market place demand, sorting out those wanted by subscribers from those rejected, and demonstrated that subscribers have a voracious appetite for programming. Cable's competitors merely have to supply alternate infrastructure and attempt to attract customers to a proven business model. The major risks have already been taken.

HISTORICAL PERSPECTIVE

Cable television is an important part of the way in which the citizens of the United States are informed and entertained. It is a means of providing large numbers of television channels to more than two thirds of the U.S. population in a cost-effective way.

The cable television system was not originally intended to be a general purpose communications mechanism. Only recently has cable's role expanded to include more general purpose communications services in some cable systems. But even in those cases, the new services are built, as much as possible, on cable's older foundations. Cable's primary and often sole purpose is the transportation of entertainment television signals to subscribers. In those applications, it needs only to be a one-way transmission path from a central location, called a headend, to each subscriber's home, delivering essentially the same signals to each subscriber. The signals are intended for use with the consumer electronics equipment which subscribers already own. This equipment is built to operate on the current U.S. television technical standard called NTSC after the organization that created it in 1941, the National Television Systems Committee. This black-and-white television standard was modified in 1953 to provide compatible color information for color television receivers and again in 1984 to add compatible stereo sound. The original purpose for cable television was to deliver broadcast signals in areas where they were not received in an acceptable manner with an antenna. These systems were called Community Antenna Television, or CATV. In 1948, Ed Parson of Astoria, Oregon, built the first CATV system consisting of twin-lead transmission wire strung from housetop to housetop. In 1950, Bob Tarlton built a system in Lansford, Pennsylvania, using coaxial cable on utility poles under a franchise from the city. Martin Malarkey opened his cable system in Pottsville, Pennsylvania, almost simultaneously (within a week of Tarlton). Malarkey used Radio Corporation of America, RCA, Antennaplex equipment which was originally developed for the in-house distribution of the National Broadcasting Company's, NBC's, signals in the RCA building in New York. Tarlton used Jerrold Corporation's antenna booster amplifier. Malarkey is a pioneer who organized and developed the National Community Television Association, NCTA which later was renamed the National Cable Television Association and more recently renamed again as the National Cable Telecommunications Association.

In most of the original cable systems, viewers found that off-air signals were either not available or were very weak because of the terrain or the distance from television transmitters. In some areas, such as New York City, multiple signal reflections and shadows cast by buildings made reception difficult. In both of these environments, a hard-wire method of delivery of signals to subscribers was welcomed. The first operators of these systems were entrepreneurial broadcasters or retail TV receiver dealers who sought to expand the market for the sale of their products by also providing the signals the products required. By the late 1960s, nearly all of the areas of the U.S. which could benefit from a community antenna had been served.

In the mid 1970s, an embryonic technology breathed new life into cable television. This technology was the satellite delivery of signals to cable systems. The pioneering movie channel, Home Box Office, HBO, showed the way.

HBO's dramatic relay of the Ali-Frazier heavyweight prize fight in October 1975 from Manila to cable systems in Mississippi and Florida effectively triggered the development of the satellite network industry. While satellites and earth stations were very expensive investments, these programming pioneers understood that the costs could be spread over many cable operators who, in turn, serve many subscribers.

Three categories of signals came into existence:

1. "Super stations" – local stations which are distributed nationally over satellite and became mini-networks. (The Turner Broadcasting System of Atlanta, Georgia, pioneered the concept.)
2. Specialized channels for news, sports, weather, education, shopping, etc.
3. Movie channels such as HBO, sparked new excitement in the business.

Cable television became much more than just a community antenna for areas with poor reception. Cable television became a means of receiving programming otherwise unavailable from broadcasters.

At present, subscribers are offered a variety of video services. The foundation service all subscribers are required to take is called basic. Off-air channels, some distant channels, and some satellite-delivered programs are included. The satellite programs include the super stations and some of the specialty channels. Pay television constitutes premium channels, usually with movies and some special events, which are offered as optional channels for an extra monthly fee. Some cable systems offer Pay-Per-View, PPV, programming which is marketed on a program-by-program basis. Recent movies and special sports events are the mainstay of PPV programming. Impulse Pay-Per-View, IPPV, allows the subscriber to order the program spontaneously, even after it has begun. The ordering mechanism usually involves two-way cable or, occasionally, an automated telephone link.

Ways of providing conditional access to allow for a limited selection of service packages at differing price points are usually included in the cable system. Simple filters remove the unsubscribed channels in some systems, while elaborate video and audio scrambling mechanisms involving encryption are used in other cable systems.

A wide variety of other services has been repeatedly offered to cable subscribers: videotext, teletext, other forms of "electronic publishing," and "information age" services, home security, and digital audio programming. Subscribers to date have shown a remarkable lack of interest in these services. It is interesting to speculate on the reasons for this disinterest, but the negative experience has been without exception. While subscribers have been very enthusiastic about video, particularly entertainment-oriented video, other services have left them cold. This has caused cable operators to concentrate their cable system design efforts on efficient, cost-effective video delivery.

Since cable television systems must utilize the public right-of-way to install their cables, they, like power, telephone, and gas companies, must obtain a franchise from the local governmental authorities. This is a non-exclusive franchise. However, experience teaches that the economics of the business generally only supports one system per community.

Spectrum Re-use

Compared with nearly any other communications technology, video is a bandwidth hog. While telephone-quality voice needs only 4 kHz of spectrum and high-fidelity sound takes 20 kHz or so (40 kHz for stereo), the current baseband video standard in the United States consumes 4.2 MHz. When modulated, the television signal takes up 6 MHz of spectrum. High definition television, HDTV, requires about 30 MHz for each of the red, green, and blue signals that make up a color picture. Extensive bandwidth compression techniques reduce the amount of spectrum required by HDTV to more acceptable amounts. These signals must then be modulated onto carriers to deliver multiple signals to the consumer's equipment.

For NTSC, each television channel is assigned 6 MHz because of the need to share limited spectrum with other services. With double sideband modulation, the 6 MHz straitjacket allows only about 2.5 MHz for the demodulated video, after allowing room for the sound carrier. In 1941, Vestigial Side Band Amplitude Modulation, VSB-AM, was invented. This modulation approach enabled television engineers to achieve higher resolution within the 6 MHz limit. Compared with double side-band amplitude modulation, VSB-AM transmits one complete side-band and only a vestige of the other. At the time the standard was created, the design requirements of practical filters determined the amount of side-band included. The consumer's receiver selects the channel to be watched by tuning a 6 MHz portion of the assigned spectrum.

In the terrestrial broadcast environment, channels must be carefully assigned to prevent interference with each other. The result of this process is that most of the terrestrial broadcast television spectrum is vacant. Better television antennas and better television circuits would allow more of the spectrum to be utilized. However, with more than 300 million receivers and more than 200 million VCRs in consumers' hands, the changeover process to upgraded systems would be difficult, costly, and require something like 20 years. Consumers continue to purchase about twenty-five million color receivers and about fifteen million VCRs per year. The reliability of the products is phenomenal. Their "half life" is about ten to twelve years.

The rest of the terrestrial spectrum which is not assigned to broadcast has other important uses. These include aircraft navigation and communications, emergency communications, and commercial and military applications. The terrestrial spectrum is too limited to supply the video needs of the U.S. viewer.

Cable television is made possible by the technology of coaxial cable. Rigid coaxial cable has a solid aluminum outer tube and a center conductor of copper-clad aluminum. Flexible coaxial cable's outer conductor is a combination of metal foil and braided wire, with a copper-clad, steel center conductor. The characteristic impedance of the coaxial cable used in cable television practice is 75 ohms. The well-known principles of transmission line theory apply fully to cable television technology.

The most important characteristic of coaxial cable is its ability to contain a separate frequency spectrum. The properties of that separate spectrum allow it to behave like over-the-air spectrum. This means that a television receiver connected to a cable signal will behave as it does when connected to an antenna. A television set owner can become a cable subscriber without an additional expenditure on consumer electronics equipment. The subscriber can also cancel the subscription and not be left with useless hardware. This ease of entry and exit from an optional video service is a fundamental part of cable's appeal to subscribers.

Since the cable spectrum is tightly sealed inside an aluminum environment (the coax cable), a properly installed and maintained cable system can use frequencies assigned for other purposes in the over-the-air environment. This usage takes place without causing interference to these other applications or without having them cause interference to the cable service. New spectrum is "created" inside the cable by the "re-use" of those frequencies. In some cable systems, dual cables bring two of these sealed spectra into the subscriber's home, with each cable containing different signals.

The principal negative of coaxial cable is its relatively high loss. Coaxial cable signal loss is a function of its diameter, dielectric construction, temperature, and operating frequency. A ball-park figure is 1 dB of loss per 100 feet. Half-inch diameter aluminum cable has around 1 dB of attenuation per 100 feet at 181 MHz; at one-inch diameter, the attenuation drops to 0.59 dB per 100 feet. The attenuation of cable varies with the square root of the frequency. Thus, the attenuation at 216 MHz (within TV channel 13) is twice that of 54 MHz (within TV channel 2) since the frequency is four times as great. If channel 2 is attenuated 10 dB in 1,000 feet, channel 13 will be attenuated 20 dB.

CABLE NETWORK DESIGN

Since cable television was originally not a general-purpose communications mechanism, but rather a specialized system for transmitting numerous television channels in a sealed spectrum, the topology or layout of the network was customized for maximum efficiency. The topology which has evolved over the years is called tree-and-branch architecture.

There are five major parts to a cable system:

1. The headend,
2. The trunk,
3. The distribution (or feeder) cable in the neighborhood,
4. The drop cable to the home and in-house wiring, and
5. The terminal equipment (set top terminals and consumer electronics hardware).

Flexible coaxial cable is used to bring the signal to the terminal equipment in the home. In the simplest cases, the terminal equipment is the television set or VCR. If the TV or VCR does not tune all the channels of interest because it is not "cable compatible," a converter is placed between the cable and the TV or VCR tuner.

Broadcast channels 2 through 13 are not in a continuous band. Other radio services occupy the gaps. Cable can re-use these frequencies because its spectrum is self-contained within the co-axial environment. The cable converter has a high-quality broadband tuner and output circuitry which puts the desired cable channel on a low-band channel not occupied in the local off-the-air spectrum. Typically this is channel 3 or 4 and occasionally 4 or 5. The TV or VCR is tuned to this channel and behaves as a monitor. If programming of interest to the subscriber is scrambled, a descrambler is required. It is usually placed in the converter. When the subscriber has purchased a product labeled "cable ready", the requirement to use a set top terminal can be a sore point.

Cable systems have been built to slightly in excess of 1 GHz with more than 150 analog channels. Cable systems with bandwidths of 750 MHz are not uncommon and 550 MHz is typical for upgrades and new construction.

The home is connected to the cable system by the flexible drop cable, typically 150 feet long. The distribution cable in the neighborhood runs past the homes of subscribers. This cable is tapped so that flexible drop cable can be connected to it and routed to the residence. The distribution cable interfaces with the trunk cable through an amplifier called a bridger amplifier, which increases the signal level for delivery to multiple homes. One or two specialized amplifiers called line extenders are included in each distribution cable. A little less than half of the system's cable footage is in the distribution portion of the plant and half is in the flexible drops to the home.

Prior to the use of fiber optics, the trunk part of the cable system transported the signals to the neighborhood. Its primary goal is to cover distance while preserving the quality of the signal in a cost-effective manner. Broadband amplifiers were required about every 2,000 feet depending on the bandwidth of the system. The maximum number of amplifiers which can be placed in a run or cascade is limited by the build up of noise and distortion. Twenty or thirty amplifiers may be cascaded in relatively high-bandwidth applications. Older cable systems with fewer channels may have as many as fifty or sixty amplifiers in cascade. Approximately 12% of a cable system's footage is in the trunk part of the system.

The headend is the origination point for signals in the cable system. It has parabolic or other appropriately shaped antennas for receiving satellite-delivered program signals, high-gain directional antennas for receiving distant TV broadcast signals, directional antennas for receiving local signals, machines for playback of taped programming and commercial insertion, and studios for local origination and community access programming.

Local origination is programming over which the cable operator has editorial control. It can range from occasional coverage of local events to a collection of programming almost indistinguishable from that of an independent broadcaster. Often, mobile coverage of events is provided with microwave links back to the headend or back-feed of the signal up the cable system to the headend. Local origination also includes alpha-numeric text as well as studio productions and video tape. In many systems, text may be the only form of local origination. Community access is programming access for community groups mandated by the franchise. The cable system typically cannot exercise editorial control over quality or content of community access programming.

Signal Quality

The ultimate goal of the cable system is to deliver pictures of adequate quality at an acceptable price while satisfying stockholders, investors, and holders of the debt generated by the huge capital expenses of building the cable system plant. This is a difficult balancing act. It would be a simple matter to deliver very high-quality video if cost were not a consideration. Experience teaches that subscriber satisfaction is a complex function of a variety of factors led by program quality and variety, reliability of service, video and sound quality and the amount of the subscriber's cable bill.

The principal picture impairments can be divided into two categories — coherent and non-coherent. Coherent impairments result in a recognizable interfering pattern or picture. They tend to be more objectionable than non-coherent impairments of equal strength.

The principal non-coherent picture impairment is noise. Random noise behavior is a well understood part of general communications theory. The familiar Boltzmann relationship, noise figure concepts, etc., apply fully to cable television technology. Random noise is the consequence of the statistical nature of the movement of electric charges in conductors. This creates a signal of its own. This noise is inescapable. If the intended analog signal is ever allowed to become weak enough to be comparable to the noise signal, it will be polluted by it yielding a snowy pattern in pictures and a sea-shore sounding background to audio.

Noise levels are expressed in cable system practice as ratios of the visual carrier to the noise in a television channel. This measure is called the Carrier-to-Noise Ratio, CNR, and is given in decibels, dB. The target value for CNR is 45 dB to 46 dB. Subjective data was compiled in 1991 in a project sponsored by the cable industry's research and development consortium Cable Television Laboratories, CableLabs. The work, supervised by Dr. Bronwyn Jones, indicates "perceptible but not annoying" CNR at about 47-50 dB. The "slightly annoying" category was within about 1 dB of 41 dB.

Coherent interference includes ingress of video signals into the cable system, reflections of the signal from transmission-line impedance discontinuities, cross modulation of video, and cross modulation of the carriers in the video signal. This latter phenomenon gives rise to patterns on the screen which are called beats. These patterns often look like moving diagonal and horizontal bars or herringbones.

The evaluation of signal quality takes place on two planes, objective and subjective. In the objective arena, measurements of electrical parameters are used. These measurements are repeatable. Standardized, automated test equipment has been developed and accepted by the video industries.

The ultimate performance evaluation involves the subjective reaction of viewers. One example of the difficulties experienced is the fact that different frequencies of noise have differing levels of irritation. High frequency noise tends to become invisible while low frequency noise creates large moving blobs which are highly objectionable. Subjective reactions to these phenomena are influenced by such factors as the age, gender, health, and attitude of the viewer. The nature of the video program, the characteristics of the viewing equipment, and the viewing conditions also impact the result.

Signal processing in the TV receiver changes the impact of signal impairments. Noise in the band of frequencies used to transmit color information is demodulated and converted into lower frequency, more objectionable noise. Noise in the synchronization part of the TV signal can cause the picture to break up entirely, resulting in much greater impairment than the same strength noise confined to other portions of the signal.

Subjective assessment of viewer reaction to picture impairments have been made using various methods. In 1959, the Television Allocations Study Organization, TASO, studied the amount of noise, interference, and distortion viewers will tolerate in a TV picture. The results were expressed in a six-point scale with grades named excellent, fine, passable, marginal, inferior, and unusable. These are very old data and the scale is no longer accepted by psychometric experts because it mixes impairment and picture quality assessments. Additionally, the lack of a mid-scale category causes difficulties in the statistical analysis. However, the broadcast industry still uses this scale even though it is considered obsolete by psychometric practitioners. The most widely accepted standard for subjective measurements of impairment is now a five-point scale which was also used in the tests sponsored by CableLabs. However, the most widely published data on subjective impact of video impairments is based on the Bell Telephone Laboratory seven-point scale not used in any other study. It is important to realize that the demand for signal quality is a function of time. Ten to 15 years ago, consumer electronics products were not capable of displaying the full resolution of the NTSC signal. Gradually, these products improved until high-end models are capable of more performance than the NTSC signal can deliver. The Super VHS and Hi-8 video tape systems have greater resolution than broadcast NTSC. As time progresses, the level of performance of consumer electronics will continue to increase. As Advanced Television, ATV, and HDTV are introduced, still more demands will be made on cable system performance. The trend to larger screen sizes also makes video impairments more evident.

CABLE SYSTEM TRADE-OFFS

The experienced cable system designer has learned how to balance noise, non-linear distortions, and cost to find a near optimal balance. Signals in cable systems are measured in dB relative to 1 mV across 75 ohms. This measure is called dBmV. This is an unfortunate name which is the source of a great deal of confusion. Since the name of the measure ends in "V", it is natural to assume it refers to a voltage. It does not. The measure, dBmV is a power ratio with reference to 13.33 nanowatts of power, defined as one millivolt, 1 mv, (rms) across 75 ohms.

Applying the well-known Boltzmann noise equation to 75-ohm cable systems yields an open-circuit voltage of 2.2 microvolts in 4 MHz at room temperature. When terminated in a matched load, the result is 1.1 microvolts. Expressed in dBmV, the minimum room-temperature noise power in a perfect cable system is -59.17 dBmV.

Starting at the home, the objective is to deliver at least 0 dBmV of signal to the terminal on the television receiver, but no more signal than the amount that would produce overload in the consumer's tuner. Lower numbers produce snowy pictures and higher numbers overload the television receiver's tuner, resulting in cross modulation of the channels. Unfortunately, the overload point of television receivers is not well defined. As a result, the FCC rules merely state that it is to be avoided.

If a converter or descrambler is used, its noise figure must be taken into account. There are two reasons for staying toward the low side of the signal range: cost and the minimization of interference in the event of a signal leak caused by a faulty connector, damaged piece of cable, or defect in the television receiver. Low signal levels may cause poor pictures when excessive signal splitting without prior amplification is used to serve too many receivers. Working our way back up the plant, we need a signal level of 10 dBmV to 15 dBmV at the highest frequency in the pass band at the tap to compensate for losses in the drop cable.

The design objectives of the distribution part of the cable system involve an adequate level of power not only to support the attenuation characteristics of the cable but to allow energy to be diverted to subscribers' premises. Energy diverted to the subscriber is lost from the distribution cable. This loss is called flat loss because it is independent of frequency. Loss in the cable itself is a square-root function of frequency and is therefore contrasted to flat loss. Because of flat losses, relatively high power levels are required in the distribution part of the plant, typically 48 dBmV at the input to the distribution plant.

Note that 0 dBm (dB with respect to 1 mW) is equivalent to 48.75 dBmV. These levels force the amplifiers in the distribution part of the plant to reach into regions of their transfer characteristics which are slightly non-linear. There is no sharply defined "nonlinear" region; it is rather a matter of degree. As a result, only a small number of amplifiers, called line extenders, can be cascaded in the distribution part of the plant. These amplifiers are spaced 300 to 900 feet apart depending on the number of taps required by the density of homes. Because the distribution part of the plant (also called the "feeder") is operated at higher power levels, non-linear effects become important. The television signal has three principal carriers, the visual carrier, the aural carrier, and the color subcarrier. The visual carrier is by far the strongest with the chroma carrier the weakest. These concentrations of energy in the frequency domain give rise to a wide range of "beats" when passed through non-linearities. To minimize these effects, the aural carrier is attenuated about 15 dB below the visual carrier.

When cable systems carried only the 12 VHF channels, second-order distortions created spectrum products which fell out of the frequency band of interest. As channels were added to fill the spectrum, second-order effects were minimized through the use of balanced, push-pull output circuits in amplifiers. The third-order component of the transfer characteristic dominates in many of these designs. The total effect of all the carriers beating against each other gives rise to an interference called Composite Triple Beat, CTB. CTB should be measured with the system fully loaded with all channels carried.

Third-order distortions increase about 6 dB for each doubling of the number of amplifiers in cascade. A 1 dB reduction in amplifier output level will generally improve CTB by 2 dB. If these products build to visible levels, diagonal lines will be seen moving through the picture. When these components fall in the part of the spectrum which conveys color information, spurious rainbows appear. The dominant non-linear effect in the feeders is the triple beat between multiple visual carriers, since the visual carriers are the points of concentration of most of the energy in the television signal.

If we assign a design level of noise and non-linear distortion at the subscriber's television receiver which is below the threshold of visibility, we can conceive of a budget of noise and distortion to be "spent" in the various parts of the system design. The distribution part of the system has relatively high powers and has used up most of the budget for non-linear distortions. On the other hand, little of the noise budget has been consumed. It can be allocated to the trunk part of the system, which brings the signals into the neighborhood.

The design objective of the trunk part of the cable system is to move the signal over substantial distances with minimal degradation. Because distances are significant, fiber and/ or lower-loss cables are used. One-inch and 0.75-inch diameter cable is common in the trunk while 0.5-inch cable is found in the distribution. Signal levels in the trunk at an amplifier's output are 30 to 32 dBmV depending on the equipment used.

It has been determined through analysis and confirmed through experience that optimum noise performance is obtained when the signal is not allowed to be attenuated more than about 20 to 22 dB before being amplified again in all-coaxial systems. Amplifiers are said to be "spaced" by 20 dB. The actual distance in feet is a function of maximum frequency carried and the cable's attenuation characteristic. High-bandwidth cable systems have their amplifiers fewer feet apart than older systems with fewer channels. But with the advent of HFC systems described below, yielding very short amplifier cascades and high channel capacity, the trend is toward spacing as high as 30 dB to 40 dB. Since attenuation varies with frequency, the spectrum in coaxial cable develops a slope. This is compensated with equalization networks in the amplifier housings.

The attenuation of the cable is a function of temperature and aging of components. Modern amplifiers use a pilot signal to regulate Automatic Gain Control, AGC, circuits. A second pilot signal at a substantially different frequency than the first allows the slope of the attenuation characteristic to be monitored and compensation to be introduced with Automatic Slope Control, ASC, circuits. Thus, long cascades of amplifiers can, once properly set up, maintain their performance over practical ranges of temperature and component aging.

Since the signal is not repeatedly tapped off in the trunk part of the system, high power levels are not required to feed splitting losses. As a result, signal levels are lower than in the distribution portion of the plant. Typical levels are about 30 dBmV. For the most part, the amplifiers of the trunk are operated within their linear regions. The principal challenge of trunk design is keeping noise under control. Each doubling of the number of amplifiers in the cascade results in a 3-dB decrease in the CNR at the end of the cascade and a 6-dB increase in the amount of CTB. If the noise at the end of the cascade is unacceptable, the choices are to employ lower noise amplifiers, shorter cascades, or a different technology such as microwave links or fiber optic links.

PERFORMANCE/COST TRADE-OFFS IN PV SYSTEMS WIRING/CABLING SELECTION

As the solar industry develops, we find many variations in design and specification requirements. Along with these variations, there is a great deal of misinformation and misunderstanding of the distinctions among various cable types and the characteristics of different cable constructions.

Additionally, with the great mission for the industry being to bring down balance of system (BoS) costs, clarity of proper applications, cable characteristics and code compliance will enable engineers to make better decisions in their choices and the performance cost trade-offs involved with different choices.

With these issues in mind we will attempt to provide some simple clarification on some of the most common issues we encounter in supplying owners, developers and contractors with the appropriate wire and cable components.

USE-2 vs PV Wire

NEC 2008, section 690.31B restricts exposed single conductor wire used to connect to PV modules and circuits to USE-2 and PV wire. Because PV modules operate at elevated temperatures and are exposed to a variety of environmental conditions, the insulation of the module interconnection must be sunlight resistant and rated for wet locations at 90oC. But what are the differences and when should each be chosen? The primary physical difference is simply a thicker insulation required on PV wire.

USE-2 is a standard wire designated for underground service entrance, typically used to connect terminals of a variety of service equipment. PV wire is specifically designated for use in PV module interconnections. Both cables are rated for 90oC wet or dry. PV wire, however, has thicker insulation to withstand the harsh environments frequently found in PV installations. USE-2 carries a 600V rating, whereas PV wire can be rated at 600V, 1000V or 2000V. There are separate ratings with increased insulation thickness for each cable.

Both USE-2 and PV wire can be used with grounded systems, but only PV wire may be used with ungrounded systems per NEC 2008. Since USE-2 has typically been installed underground or in environments where flame is unlikely, it usually does not require a flame retardant rating. The PV wire designation does include a VW-1 flame test requirement. Both ratings are sunlight resistant, but the PV wire rating applies a more stringent test requirement of 720 hours in weatherometer vs. 300 hour weatherometer test for USE-2. In summary, PV wire provides, through thicker insulation, greater protection and a wider range of suitable applications.

DIFFERENTIATING CABLE CONSTRUCTIONS

One of the biggest areas of confusion is with regard to the terms double insulation, double-jacket and dual-layer. This is an important area to clarify as a number of these terms have been used by marketing departments to enhance the perceived performance and applications for certain products. We have found that engineers using the word "double," are looking for two separate layers, which would mean an insulation layer with a separate protective jacket.

Dual-layer (also called double-pass) should be two independent layers usually consisting of different compounds such as ethylene-propylene-rubber (EPR) insulation and a chlorinated polyethylene (CPE) jacket. This combination of EPR/CPE provides the highest level of weather/sunlight resistance and electrical properties. Single-pass comprises one insulation layer, such as an XLPE or EPR compound.

The primary compounds used for PV wire include cross-linked polyethylene (XLPE), cross-linked polyolefin (XLPO), which is also a low smoke zero-halogen product; ethylene-propylene-rubber (EPR), and chlorinated polyethylene (CPE). XLPO has been used extensively in the

European market. However, it's main benefit, low smoke zero-halogen, is really designed for human safety in enclosed environments in the case of fire, such as in a tunnel.

Since PV installations are outdoors in open spaces, there is no reason to pay the substantial increase in price for an attribute that will realize no benefit in PV installations. Polyvinyl chloride (PVC) is not included in the list because as a thermoplastic compound it is subject to softening or melting when exposed to high temperatures. Single pass XLPE and a dual layer EPR/CPE are the two most common constructions in the North American market. An EPR/CPE combination provides better protection against UV, ozone, heat, abrasion and oxidation. Additionally, separate layers of insulation and jacket means that if the cable is subject to a nick or abrasion, the insulating layer can remain uncompromised, thus retaining the integrity of electrical insulation.

Fine-strand vs Class B-strand Conductors

Many module manufacturers, as well as many plant installations, particularly in Europe, use a fine-stranded copper conductor rather than the standard class B 7 strand wire. The fine stranding provides a more flexible cable, which is easier to pull through conduit and has tighter bend radius. Most US installations specify 7-strand class B conductor. The questions come mostly from installers or engineers who are picking up a European design.

John Wiles, of the Southwest Technology Development Institute, New Mexico State University, wrote a fine article addressing these issues. First of all, class B 7 strand is the standard conductor stocked by most manufacturers and distrbutors in North America, so it tends to be more readily available and less expensive. Beyond that, the most compelling reason for specifying 7 strand over fine strand is the integrity of terminations at the combiner box.

Wiles points out that there have been reports of failed connections when fine-stranded conductors are used with mechanical terminals or lugs that use a set screw to hold the wire in the terminal. The problem is threefold. First, the screw has a tendency to break some of the fine strands of the conductor. Second, some of the fine strands can fray and not fully connect. Third, the fine-stranded conductors will continue to compress after the initial torque setting, eventually resulting in a loose connection. Subsequent re-torqueing can result in a higher-than-normal resistance, which heats the copper and contact, causing expansion and further loosening. Ultimately, there is an increased risk of failure with fine strand if not used with appropriate terminations.

UL engineers have said that few (if any) of the normal screw type mechanical terminals that the PV industry commonly uses have been listed for use with fine-stranded wires. Even if the modules comes with a fine-stranded wire, the end of string or "home-run" cable that connects the modules or string to a combiner box, typically will not require fine-stranding.

Direct Burial Cables

PV wire with direct burial rating or USE-2 cable, is typically not required to be installed in conduit. The slight increase in cost for these cable types vs. using the less expensive THHN/THWN-2 in conduit provides far greater savings vs. the additional costs of conduit materials and additional labor required to assemble and install conduit systems. In the interest of bringing down BoS costs, more consideration should be given to developing designs with direct burial cable.

CAREER AS A CABLE TELEVISION AND TELECOMMUNICATIONS TECHNICIAN

DEFINITION AND NATURE OF THE WORK

Cable television and telecommunications technicians install, maintain, and upgrade cable systems. They also perform repair work and respond to service problems at the homes of cable users. Their main goal is to keep a cable system operating efficiently.

Cable technicians generally start out as cable system installers. Installers prepare a subscriber's home for cable reception by connecting the cable system to one or more television sets or to one or more computers for high-speed Internet service. Installers explain to subscribers how to operate the cable system and how they can upgrade or downgrade their services. If a customer no longer wants the cable system, installers remove it. Installers may also be responsible for servicing feeder lines, which connect the main cable to small groups of homes, and droplines, which are the direct cable lines to a subscriber's home. In general, installers perform light troubleshooting work when customers have minor problems with their cable units.

Trunk technicians must have more advanced technical skills than installers. Trunk technicians continuously maintain the main line, or trunk line, of the cable system. This work is very important because the trunk line is the main artery of a cable system and a malfunction could shut down service in a large area. Trunk technicians are responsible for correcting any technical failure in the trunk line, the feeder amplifier, or the booster.

Service technicians respond to problems reported by a subscriber. These technicians generally repair amplifiers, cable poles, or lines. They may also correct electrical malfunctions occurring in the cable system. They practice preventive maintenance by electronically scanning the cable system periodically. By catching minor problems early, they can prevent major disruptions in cable service later.

Not all cable television and telecommunications technicians work in the field. Bench technicians, who are highly specialized electronics technicians, operate the cable system's repair facility. They examine broken and malfunctioning cable equipment, repair it, record the repairs, and return the equipment to the field. Chief cable technicians are the most highly skilled of the cable technician staff. They supervise the work of other technicians. They are ultimately responsible for ensuring high-quality signal delivery to customers. Chief cable technicians also set employee performance standards, conduct interviews, and handle personnel matters. They do not work out in the field unless serious or complex problems arise.

Education and Training Requirements

Cable television and telecommunications technicians need at least a high school education, an aptitude for math, and an interest in installing or repairing electronic equipment.

Most cable television and telecommunications technicians begin their careers as installers. Training is available for high school graduates at trade schools or from a cable company's on-the-job training program. Some trade schools offer a five- to seven-week course in electronics repair and maintenance. Trunk, service, and bench technicians have specialized telecommunications training; junior colleges and technical institutes offer programs that usually last one to two years. Chief cable television and telecommunications technicians need at least two years of electronics training and a considerable amount of experience in the field.

The number of vocational schools that train cable television and telecommunications technicians is limited. The cable industry provides on-the-job training, as well as special training schools for technicians. Some community agencies and educational institutions also offer technical training programs for cable technicians.

Getting the Job

Schools that offer courses in cable maintenance often have placement offices that can help candidates find jobs. The majority of employment opportunities come from areas that already have cable service or are about to get it. The National Cable and Telecommunications Association (NCTA) provides lists of these areas. State and local government officials can help prospective cable technicians determine what opportunities exist locally.

The Television and Cable Factbook, published by Warren Communications, lists all the cable systems in the United States and Canada. Copies of the print edition can be found at most large libraries. Also available online, The Factbook is considered the industry's leading source of up-to-date information.

Advancement Possibilities and Employment Outlook

Cable television technicians generally advance by gaining experience in cable maintenance and servicing. Installers, trunk technicians, service technicians, and bench technicians can advance to become chief cable technicians with increased on-the-job experience or electronics training. Some technicians may elect to continue their college training and become professional engineers. Those who have the necessary teaching skills may train new cable television and telecommunications technicians.

The employment outlook for cable technicians is slower than the average through the year 2014. Most areas of the country that can be served by cable systems have already been wired, and fewer installations will be needed in the coming years. However, technicians will still be needed to service existing cable lines and to replace old wiring with fiber optic cable. Steady customer demand for high-speed Internet connections may increase the demand for cable technicians; however, the increasing use of wireless Internet systems, along with the option of satellite television, will most likely counteract any job growth in the cable television and telecommunications field. Jobs will become available as workers retire or leave their jobs for other reasons.

Working Conditions

Installers, service technicians, and trunk technicians spend most of their time outside installing or maintaining cable lines, while bench technicians perform their tasks inside the cable system's repair facility. Chief technicians normally do not go into the field unless a problem arises. All technicians in this field should enjoy working with the public because it is a consumer-oriented industry. Cable television technicians generally work five days per week, but they may have to work some overtime.

Earnings and Benefits

The earnings of cable technicians depend on the worker's experience, education, and place of employment. According to the U.S. Bureau of Labor Statistics, the median salary of a cable television and telecommunications technician is $17.36 per hour. Chief cable technicians earn more. Benefits include paid holidays and vacations, health insurance, and retirement plans. Cable technicians also receive extra pay for overtime.

THE HYBRID FIBER COAX ARCHITECTURE

The Hybrid Fiber Coax, HFC, approach breaks the cable system into a multitude of much smaller cable systems with amplifier cascades limited to four to six amplifiers. Each of these small cable systems is fed with a fiber link to the headend. Surprisingly, the trunk portion of the cable plant only covers about twelve percent of the total footage. The feeder portion of the cable system supports taps for subscribers. It is usually constructed to no more than a mile and a half in length. This limitation comes from the fact that energy is tapped off to feed homes. Consequently, the power levels must be relatively high. These higher power levels reach into the slightly non-linear regions of the amplifiers. As a result, only a few amplifiers can be used before the distortion begins to degrade the picture quality. Approximately 38% of the total footage in a cable system is in the feeder portion of the plant. The drop is the flexible cable which goes to the home. It is usually constructed to no more than 400 feet in length. Typically, it is more like 150 to 200 feet long. Approximately half of the total footage of cable in a system is in the drop and the flexible wiring in the home.

The feeder portion of the cable plant is a hotbed of activity. Every day, new subscribers are added and others are removed. Approximately twenty percent of Americans move to a new residence every year. When they move, the drop is usually disconnected. When they move to their new residence, that drop must be re-connected. It is important that the technology used in the feeder portion of the trunk supports this constant activity. It must be "craft-friendly". That is, it must be easy to work with. The trunk portion of the cable plant is relatively stable. There, few changes are made.

The HFC architecture is an optimized combination of fiber in the trunk and coaxial cable in the feeder and drop. HFC has made it possible to cost effectively increase bandwidth, signal quality, and reliability, while reducing maintenance costs and retaining a craft-friendly plant. It makes two-way service practical.

The concepts behind HFC are straight forward. The bandwidth of coaxial cable has no sharp cut-off. It is the cascade of amplifiers that limits the system bandwidth. Twenty to forty amplifiers in cascade not only reduces bandwidth, but also constitutes a severe reliability hazard. Overlaying low-loss fiber over the trunk portion of the plant eliminates or reduces the number of trunk amplifiers. Wider bandwidth is thus facilitated. Two-way cable operation becomes practical because of two benefits of HFC. First, the cable system is broken up into a large number of smaller cable systems, each isolated from the others by its own fiber link to the headend. If ingress should cause interference in one of these small cable systems, that interference will not impair the performance of the other portions of the cable plant. Secondly, the fiber itself is not subject to ingress of interfering signals.

Today it is common practice to install passive components (taps, splitters, amplifier housings) that will pass 1 GHz. Actual 1-GHz amplifiers are still too costly. However, some systems are being build "GHz ready" so that the addition of GHz amplifier modules will upgrade the system with little or no waste of installed plant. While GHz capability is a future possibility, 750 MHz upgrades are done today and 550 MHz is commonplace.

The lasers that drive the fibers are expensive, costing many thousands of dollars. The receivers that convert the optical energy into signals which TVs and VCRs can use are also modestly expensive. To be practical, these components must serve hundreds of subscribers each so that the costs can be shared. This is accomplished by fibers that feed small coaxial systems, which in turn serve a few hundred to a couple of thousand subscribers.

A major attraction of HFC is that its implementation cost is low. Fiber is "over lashed" onto the existing trunk plant. The in-place cable is broken into segments and used for the small-scale cable systems. Some of the amplifiers are reversed in direction. Nothing is wasted.

SIGNAL SECURITY SYSTEMS

Means exist for securing services from unauthorized viewership. These range from simple filtering schemes to remote controlled converter/ descramblers. The filtering method is a commonly used method of signal security and is the least expensive.

Trapping Systems

There are two types of filtering or trapping schemes: positive trapping and negative trapping. In the positive trapping method, an interfering jamming carrier (or a set of multiple carriers) is inserted into the video channel at the headend. If the customer subscribes to the secured service, a positive trap is installed at the customer's house to remove the interfering carrier. The positive trapping scheme is the least expensive means of securing a channel in systems where less than half of the basic customers subscribe to the premium service.

A drawback to positive trap technology is its defeatability by customers who obtain their own filters through theft, illegal purchase, or construction. Another drawback is the loss of resolution in the secured channel's video caused by the filter's effect in the center of the video passband. Pre-emphasis is added at the headend to correct for the filter's response, but loss of picture content in the 2-3 MHz region of the baseband video signal remains. New positive trap schemes take advantage of sharper Surface Acoustic Wave, SAW, or crystal filter technology. This approach allows the interfering carriers to be positioned a few kilohertz away from the secured channel's visual carrier in contrast to 2.25-2.75 MHz with the conventional approach. Illegal construction of these filters is much more difficult. The new technology has the promise of achieving greater signal security and a higher quality picture for the authorized customer, but at a higher price.

Negative trapping removes signals from the cable drop to the customer's home. The trap is needed for customers who do not subscribe. This is the least expensive means of securing a channel in systems where over half of the basic customers subscribe to the premium service. The negative trap is ideal. There is no picture degradation of the secured channel because the trap is not in the line for customers who take the service. A drawback occurs for customers who do not subscribe to the secured service but want to view adjacent channels. These customers may find a slightly degraded picture on the adjacent channels due to the filter trapping out more than just the secured channel. This problem becomes more significant at higher frequencies, due to the higher Q (efficiency) required of the filter circuitry. Negative traps have a problem from a security standpoint. The customer merely has to remove the negative trap from the line to steal an unauthorized service. Maintaining signal security in negative trapped systems depends upon ensuring that the traps remain in the drop lines.

Scrambling and Addressability

There are two classes of scrambling technologies:

1. RF synchronization suppression systems and
2. Baseband scrambling systems.

The concept of addressability should be considered separately from the scrambling method. Non-addressable converter/descramblers are programmed via internal jumpers or a semiconductor memory chip called a prom (programmable read only memory) to decode the authorized channels. These terminals' authorizations must be physically changed by the cable operator. Addressable converters are controlled by a computer-generated data signal originating at the headend and placed either in the vertical blanking interval, VBI, or in a separate RF carrier. This signal remotely configures the viewing capabilities of the converter. Impulse-pay-per-view, IPPV, technology is supported by addressable converter/descrambler systems.

RF Synchronization Suppression Systems

Converter-based scrambling systems that perform encoding and decoding of a secured channel in an RF format comprise the commonly used scrambling technology. There are two basic RF scrambling formats. The more common is known as gated or pulsed synchronization suppression. With this method, the horizontal synchronizing pulses (and with some manufacturers, the vertical synchronization pulses) are suppressed by 6 dB and/or 10 dB. This is done in the channel's video modulator at the IF frequency. The descrambling process in the converter/descrambler occurs at its channel output frequency. This is accomplished by restoring the RF carrier level to its original point during the horizontal synchronization period. Variations of this method pseudo-randomly change the depth of suppression from 6 dB to 10 dB or only randomly perform suppression.

The older, and now less popular, format is known as sine-wave synchronization suppression. In this format, the scrambling effect is achieved by modulating the visual carrier signal with a sine wave. This causes the synchronization to be suppressed, as well as changing the characteristic content of the basic video information. This encoding process is performed at the IF frequency in the channel's video modulator at the headend. The decoding process in the converter/descrambler is accomplished by modulating the secured channel with a sine wave of the same amplitude and frequency as in the headend, but whose phase is reversed.

A phase-modulated RF scrambling technique based on precision matching of SAW filters constructed on the same substrate has been introduced. This low-cost system has extended operators' interest in RF scrambling techniques for use within addressable plants.

Baseband Scrambling Systems

Baseband converter/descrambler technology provides a more secure scrambling technology for delivering video services. The common encoding format is a combination of random or pseudo-random synchronization suppression and/or video inversion. More complex and effective schemes have been invented but have not found broad acceptance because of cost. Because the encoding and decoding are performed at baseband, these converter/descramblers are complex and expensive.

Maintenance of the system's video quality is an ongoing issue. The encoders are modified video processing amplifiers. They provide controls to uniquely adjust different facets of the video signal. The potential for set-up error in the encoder, in addition to the tight tolerances that must be maintained in the decoders, has presented challenges to the cable operator.

A recent extension to baseband video scrambling involves the addition of digitized and encrypted audio to the video signal's vertical and horizontal synchronization pulse intervals. While there are issues of stereo compatibility, providers of these products claim that the combination of video scrambling with audio encryption provides a significantly higher degree of security.

Off-Premises Systems

The off-premises approach is compatible with recent industry trends to become more consumer electronics friendly and it removes security-sensitive electronics from the customer's home. This method controls the signals at the pole rather than at a decoder in the home. This increases consumer electronics compatibility since authorized signals are present in a descrambled format on the customer's drop. Customers with cable-compatible equipment can connect directly to the cable drop without the need for converter/descramblers. This allows the use of all VCR and TV features.

Signal security and control in the off-premises devices take different forms. While there were several attempts to take modified addressable converter/descramblers and enclose them on the pole, this approach was not successful. The system was costly and less consumer friendly than having the converter/descrambler in the home because it delivered only a single channel at a time to the customer's equipment.

Interdiction technology involves a scheme similar to that of positive trap technology. In this format, the pay television channels to be secured are transported through the cable plant in the clear (not scrambled). The security is generated on the pole at the subscriber module by adding interference carrier(s) to the unauthorized channels. An electronic switch is incorporated allowing signals to be turned off. In addition to being consumer electronics friendly, this method of security does not degrade the picture quality on an authorized channel.

The Signal and the Customer's Equipment

The common devices at the cable drop to the customer's home are grounding safety devices called ground blocks and a two-way signal splitter that sometimes has a built-in grounding terminal.

Some systems use ground blocks or two-way splitters that incorporate a high-pass filter. These filters are used in two-way plant to minimize RF ingress into the cable plant in the 5-30 MHz reverse (upstream) spectrum from noise sources in the home. These noise sources include appliances with motors, and consumer electronics products with spurious emissions in these frequencies. These filters have a low enough crossover frequency not to cause group delay in the downstream video channels.

Splitters or ground blocks should have little effect on picture quality provided there is adequate signal to handle the splitter's loss. The signal strength may be below specifications due to an excessively long drop or more activated cable outlets in the house than the cable design anticipated. To compensate, some systems use an AC-powered drop amplifier. These amplifiers can create problems — a reduced CNR or increased distortions.

Consumer electronics switching devices, designed to allow convenient control and routing of signals between customer products and cable systems' converters/descramblers, have built-in amplification stages to overcome the losses associated with the internal splitters. These amplifiers add distortions or noise. When cable systems were designed, consumer electronics switching devices were not taken into account because they did not exist.

Signal splitting in VCRs can be a problem. To compensate for recording SNR deficiencies, low grade VCRs sometimes split the signal unequally between the by-pass route and the VCR tuner. This gives the VCR a stronger signal than the TV receiver to improve VCR performance. In addition, this strategy reduces the quality of the signal routed to the TV. When it is compared with VCR play-back, the disparity in performance is reduced.

Consumer Electronics Compatibility

In the early 1980s, a Joint Engineering Committee, JEC, was formed under the sponsorship of the Electronic Industries Association, EIA, and the NCTA. The purpose of the JEC was to bring together technical representatives from the cable and the consumer electronics industries to attempt to find ways to make cable service and consumer electronics products more compatible. It was soon recognized that the cable subscriber and the consumer electronics customer were one and the same. In order to better serve that customer, cooperation between the two industries was necessary. The two industries had for decades evolved slowly, but without coordination. As technology began to accelerate, a serious divergence became apparent. It was the consumer who ultimately suffered. The high-minded goal of improving the situation for the consumer ran into a few practical problems.

Problems on the consumer side stem from critical cost considerations. Additional expense is hard to recover by raising prices because of the extreme competitiveness of that industry. That competitiveness is a consequence of the ability to produce larger quantities of very reliable, high quality products than the marketplace can easily absorb. This over-capacity makes prices almost impossible to sustain, much less raise. The nature of technology is such that it is impossible to contain it. Any technical advance quickly spreads throughout the industry making differences between most brands difficult for the consumer to distinguish.

Problems on the cable side come from the need for signal security and the desire to experiment with new services. A prerequisite to offering a subscription service is the ability to deny that service to those who choose not to pay. Without payment, programming cannot be purchased and those who create that programming cannot be compensated. So signal security is critical to the delivery of subscription services. Technology makes possible experimentation in new services. Because these services are offered in set-top terminals owned by the cable operator, the consumer is protected from financial loss if service fails.

Early on, it was clear that three things were needed:

1. A standardized channel plan vs. frequency
2. Tuner Specifications
3. A Decoder Interface

With these three elements, much of the difficulty could be resolved. A major feature of cable television is its self-contained spectrum. Because this spectrum is isolated from the spectrum in the environment, the usual restrictions placed on broadcast television do not apply. The entire spectrum can be filled with television signals. Early cable practice utilized a variety of methods for labeling these channels. This diversity made it difficult to use a common labeling scheme for consumer electronics products with extended tuning range. A frequency plan was necessary to allow common nomenclature.

These cable signals can be very well controlled in their relative strength. This is in sharp contrast to broadcast signals. To prevent broadcast co-channel interference, channels are spaced. The VHF and UHF bands have gaps where others have claim on the spectrum. It is common for the broadcast tuner to attempt to tune a distant weak signal centered between nearly adjacent strong local signals. Thus the demands on tuners designed for broadcast and for cable are very different. It is difficult to make a tuner which performs well in both situations.

A most serious difference in the requirements of tuners designed for cable and those designed for broadcast is the issue of direct pick up. A broadcast tuner needs minimal electromagnetic shielding to protect it from outside signals. The spectrum being tuned is only occupied by the broadcast signal. A cable tuner has another problem. Because the propagation speed through cable is about two thirds that of free space, the broadcast signal arrives at the TV first. A few microseconds later, the cable signal arrives. If the tuner directly responds to the over-the-air signal, the two signals will be mixed and interference will result. The solution to this problem is covered by a now-expired U.S. Patent, the Mandell patent. The cable signal is selected by a tuner specifically designed for the purpose. It is well shielded and has the ability to discriminate against immediately adjacent channels. One constraint is relaxed. Signal levels across the tuned band differ only slightly in level. This tuner's output is converted to a single channel chosen to be one not used locally. Therefore, there is no off-air signal to be directly picked up by the TV and mixed with that delivered by the cable system. The problem of direct pick up is solved only because the output of the converter is a single channel on a frequency not used locally.

The use of the set-top converter solves one problem but makes the TV's tuner useless. Those features which depend on the tuner are also made inoperable. In particular, the remote control that came with the TV or VCR is only used to turn those devices on and off. The set-top's remote control changes channels. In the case of the VCR, the timer which changes channels is crippled because it cannot change the channels of the converter. Taping different channels consecutively becomes impossible.

In order for a "cable ready" TV or VCR to provide acceptable performance when connected to cable, it has to have a minimum amount of electromagnetic shielding to make it immune to direct pick up interference. Also, it must not radiate spurious signals up the cable. These signals can come from the local oscillator or from the harmonics of all of the digital signals which provide on-screen displays and drive other features. These signals can cause interference to other receivers in the same residence or in adjacent residences if they are allowed to enter the cable. All of these issues have been covered by a Cable RF specification.

Cable's requirement for signal security caused the need for set-top terminals to move beyond mere frequency conversion. Set top terminals took on the function of descrambling signals in the early 1980's. If a "cable ready" TV or VCR were designed with adequate tuning range, adequate electromagnetic shielding against direct pick-up, and the ability to handle a full spectrum of adjacent signals without overload, it might still require a set-top descrambler. This would duplicate the circuitry in the TV, provide slight additional signal degradation, and interfere with certain features built into TVs and VCRs. The solution to this problem is the Decoder Interface connector.

This is a twenty-six pin connector on the back of TVs and VCRs which brings video and audio signals out of the consumer electronics product for external processing. The descrambled signals are returned to the consumer electronics product for display or recording. The main advantage of the Decoder Interface is that it makes scrambling nearly transparent to the user.

In 1992, Congress passed the Cable Act which has far-reaching implications for cable system operation and economics. An amendment attached to the Cable Act requires the FCC to create rules with the intent of improving the compatibility between consumer electronics products and cable operations. These rules regulate the conditions under which scrambling can be employed and provide a technical definition for "cable ready" products. That definition included the channel plan created by the JEC, specifications on tuner performance, and the inclusion of a Decoder Interface connector. Products which do not comply with the Commission's technical specifications cannot be sold as "cable ready", "cable compatible", or under any other term which might mislead consumers.

As complex and difficult as the Decoder Interface has been, the digital television future presents even more interesting challenges. Digital Television, DTV, takes advantage of the fact that we can now afford to put millions of digital transistors into consumer products to process signals. Additional millions of transistors can be included in memory elements. This tremendous processing power offers a wide variety of options for digital video compression and new interactive and Multi-Media services. The biggest unknowns center on the consumer. Which services will find acceptance? How much tolerance exists for the new and different digital artifacts that this new signal processing brings?

The 1996 Telecommunications Act requires the FCC to establish rules for the availability of set top terminals at retail. As we move to subscriber ownership of the electronics which implement services, new issues become important. In the old paradigm, the service provider answered questions and supplied training in the use of the equipment. The equipment was repaired at no additional cost. If the service failed to enjoy sustaining support, the cable operator absorbed the equipment loses. In the new world of subscriber ownership, the retailer and manufacturer of the equipment must stand ready to answer questions on usage and provide training and repair. The consumers' expectations must be adjusted to accept situations in which the equipment becomes no longer useful because the service failed or was replaced with a improved version. The cable operator will no longer be absorbing the losses since the equipment is no longer his.

An additional complication arises. If subscriber-owned set-top terminals introduce interfering signals into the cable system, they will have to be disconnected. If the subscriber insists on using them, that subscriber will have to be disconnected. There will be charges for service calls to solve these problems. Consumer will be angry when the hardware they purchase cannot be used on the cable system because of interference it may cause.

DIGITAL VIDEO COMPRESSION

Digital Video Compression, DVC, has given massive capacity to the cable spectrum. DVC gained its first momentum in the pursuit of HDTV. The analog bandwidth of signals from an HDTV camera consists of something like 30 MHz each of Red, Blue, and Green. Nearly 100 MHz of analog signals would require more than a gigabit per second for straight digital transmission. The FCC, which sets federal regulations on spectrum usage, only allows 6 MHz! As a result, there is a need to remove a lot of the redundancy in the picture. The same technology can be used to squeeze multiple standard definition signals into the same 6 MHz. When DVC is used to put multiple current quality signals into 6 MHz, the result is called Standard Definition Television, SDTV.

DVC was first used by consumers in the Direct Broadcast Satellite, DBS, environment. DirecTV, EchoStar, and the former PrimeStar use(d) a version of the Motion Picture Experts Group, MPEG, standard. DVC compresses signals by removing redundancy in each still picture, by removing redundancy between the series of still pictures that make up a motion sequence, and by deleting those aspects of the image which have minimal visibility to the human visual system.

The reasons for the lower data rate with movies stems from two main factors. First, movies have a twenty-four frames per second rate while video has thirty frames per second. While motion reproduction in video is better than in movies, there is a twenty percent increase in data required for the same image quality. Secondly, movies can be processed iteratively. The compressor parameters can be adjusted on a scene by scene basis. Much more processing can be brought to bear since the computations do not have to be done "in real time".

The other factor which determines the packing density of movies in the 6 MHz television channel is the modulation method. The two main competitors are Quadrature Amplitude Modulation, QAM, and Vestigial Side Band, VSB, modulation. These come in various data speed capacities. The two most interesting for cable applications are 256-QAM or 16-VSB. Both are "double data rate" systems in that they can deliver data payloads of about double the data rate required to deliver one HDTV signal in 6 MHz. The Advanced Television Test Center, ATTC, completed tests of 16-VSB on cable in Charlotte, North Carolina.

The 16-VSB system tested delivered 38.5 Mb/s in all locations. Some older cable systems were included in the tests. The longest amplifier cascade was forty-eight! Fiber links and microwave Amplitude Modulated Links, AMLs, were also included. 256-QAM delivers similar results. Thus it is possible to transmit twelve DVC movies in 6 MHz at 3 Mb/s each.

VIDEO COMPRESSION STANDARDS – PROS & CONS

Video Compression is the term used to define a method for reducing the data used to encode digital video content. This reduction in data translates to benefits such as smaller storage requirements and lower transmission bandwidth requirements, for a clip of video content.

Video compression typically involves an elision of information not considered critical to the viewing of the video content, and an effective video compression codec (format) is one that delivers the benefits mentioned above: without a significant degradation in the visual experience of the video content, post-compression, and without requiring significant hardware overhead to achieve the compression. Even within a particular video compression codec, there are various levels of compression that can be applied (so called profiles); and the more aggressive the compression, the greater the savings in storage space and transmission bandwidth, but the lower the quality of the compressed video [as manifested in visual artefacts – blockiness, pixelated edges, blurring, rings – that appear in the video] and the greater the computing power required.

VIDEO COMPRESSION

Video compression technologies are about reducing and removing redundant video data so that a digital video file can be effectively sent over a network and stored on computer disks. With efficient compression techniques, a significant reduction in file size can be achieved with little or no adverse effect on the visual quality. The video quality, however, can be affected if the file size is further lowered by raising the compression level for a given compression technique. Different compression technologies, both proprietary and industry standards, are available. Most network video vendors today use standard compression techniques. Standards are important in ensuring compatibility and interoperability. They are particularly relevant to video compression since video may be used for different purposes and, in some video surveillance applications, needs to be viewable many years from the recording date. By deploying standards, end users are able to pick and choose from different vendors, rather than be tied to one supplier when designing a video surveillance system.

Axis uses three different video compression standards. They are Motion JPEG, MPEG-4 Part 2 (or simply referred to as MPEG-4) and H.264. H.264 is the latest and most efficient video compression standard.

Video Codec

The process of compression involves applying an algorithm to the source video to create a compressed file that is ready for transmission or storage. To play the compressed file, an inverse algorithm is applied to produce a video that shows virtually the same content as the original source video. The time it takes to compress, send, decompress and display a file is called latency. The more advanced the compression algorithm, the higher the latency.

A pair of algorithms that works together is called a video codec (encoder/decoder). Video codecs of different standards are normally not compatible with each other; that is, video content that is compressed using one standard cannot be decompressed with a different standard. For instance, an MPEG-4 decoder will not work with an H.264 encoder. This is simply because one algorithm cannot correctly decode the output from another algorithm but it is possible to implement many different algorithms in the same software or hardware, which would then enable multiple formats to coexist.

Image Compression vs. Video Compression

Different compression standards utilize different methods of reducing data, and hence, results differ in bit rate, quality and latency. Compression algorithms fall into two types: image compression and video compression.

Image compression uses intraframe coding technology. Data is reduced within an image frame simply by removing unnecessary information that may not be noticeable to the human eye. Motion JPEG is an example of such a compression standard. Images in a Motion JPEG sequence is coded or compressed as individual JPEG images.

Fig. With the Motion JPEG Format, the Three Images in the above Sequence are Coded and sent as Separate Unique Images (I-frames) with no Dependencies on each Other.

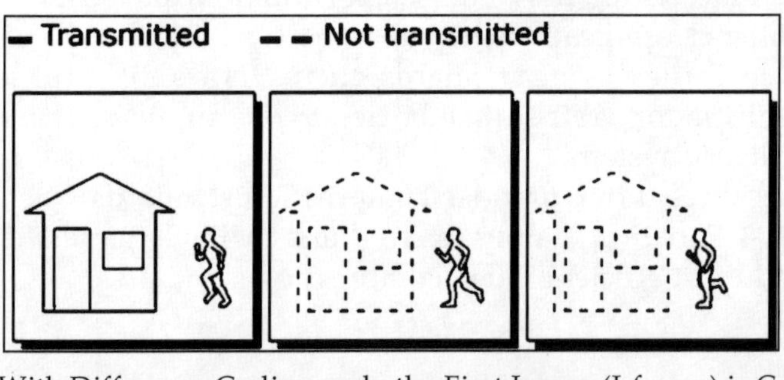

Fig. With Difference Coding, only the First Image (I-frame) is Coded in its Entirety. In the Two following Images (P-frames), References are made to the First Picture for the Static Elements, i.e. the House. Only the Moving Parts, i.e. the Running Man, are Coded using Motion Vectors, thus Reducing the Amount of Information that is Sent and Stored.

Video compression algorithms such as MPEG-4 and H.264 use interframe prediction to reduce video data between a series of frames. This involves techniques such as difference coding, where one frame is compared with a reference frame and only pixels that have changed with respect to the reference frame are coded. In this way, the number of pixel values that is coded and sent is reduced. When such an encoded sequence is displayed, the images appear as in the original video sequence.

Other techniques such as block-based motion compensation can be applied to further reduce the data. Block-based motion compensation takes into account that much of what makes up a new frame in a video sequence can be found in an earlier frame, but perhaps in a different location. This technique divides a frame into a series of macroblocks (blocks of pixels). Block by block, a new frame can be composed or 'predicted' by looking for a matching block in a reference frame. If a match is found, the encoder codes the position where the matching block is to be found in the reference frame. Coding the motion vector, as it is called, takes up fewer bits than if the actual content of a block were to be coded.

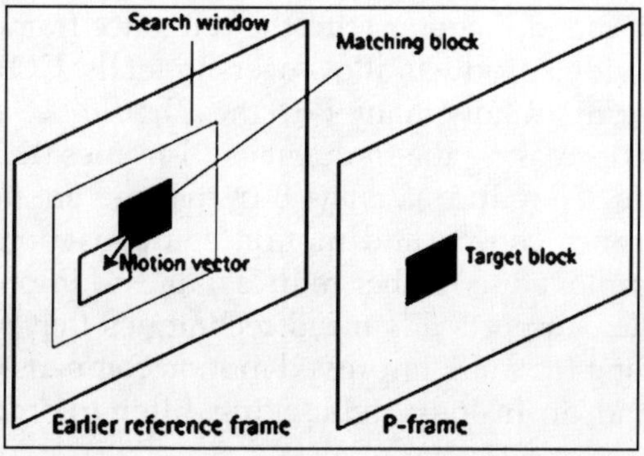

Fig. Illustration of Block-based Motion Compensation.

With interframe prediction, each frame in a sequence of images is classified as a certain type of frame, such as an I-frame, P-frame or B-frame. An I-frame, or intra frame, is a self-contained frame that can be independently decoded without any reference to other images. The first image in a video sequence is always an I-frame. I-frames are needed as starting points for new viewers or resynchronization points if the transmitted bit stream is damaged. I-frames can be used to implement fast-forward, rewind and other random access functions. An encoder will automatically insert I-frames at regular intervals or on demand if new clients are expected to join in viewing a stream. The drawback of I-frames is that they consume much more bits, but on the other hand, they do not generate many artifacts, which are caused by missing data.

A P-frame, which stands for predictive inter frame, makes references to parts of earlier I and/or P frame(s) to code the frame. P-frames usually require fewer bits than I-frames, but a drawback is that they are very sensitive to transmission errors because of the complex dependency on earlier P and/or I frames. A B-frame, or bi-predictive inter frame, is a frame that makes references to both an earlier reference frame and a future frame. Using B-frames increases latency.

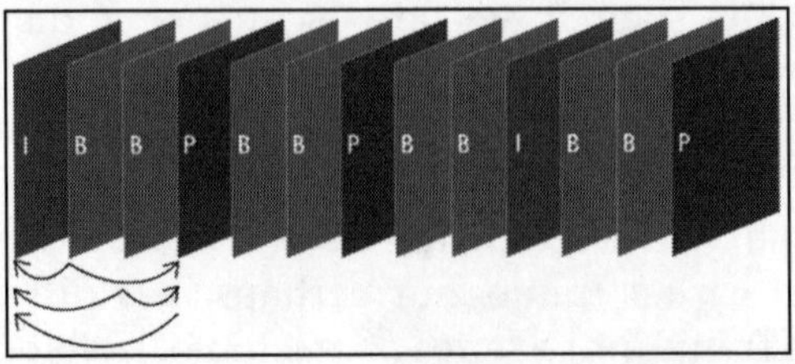

Fig. A Typical Sequence with I-, B- and P-frames. A P-frame may only Reference Preceding I- or P-frames, While a B-frame may Reference both Preceding and Succeeding I- or P-frames.

When a video decoder restores a video by decoding the bit stream frame by frame, decoding must always start with an I-frame. P-frames and B-frames, if used, must be decoded together with the reference frame(s).

Axis network video products allow users to set the GOV (group of video) length, which determines how many P-frames should be sent before another I-frame is sent. By decreasing the frequency of I-frames (longer GOV), the bit rate can be reduced. To reduce latency, B-frames are not used.

Besides difference coding and motion compensation, other advanced methods can be employed to further reduce data and improve video quality. H.264, for example, supports advanced techniques that include prediction schemes for encoding I-frames, improved motion compensation down to sub-pixel accuracy, and an in-loop deblocking filter to smooth block edges (artifacts).

3

ICTs in Technical and Vocational Education and Training

ROLE OF ICTS FOR TEACHING AND LEARNING

ICTs are revolutionizing education by removing distance from education and making knowledge more accessible to all (Industry Canada, 1997). Technology-enhanced learning will play a crucial role in the development of a lifelong learning culture and has the capacity to empower learners by providing them with multiple pathways that offer choices and channels to meet their education and training needs (Human Resources Development Canada, 1998). It is not surprising therefore to see a growing interest in Technology-Based Learning (TBL) across the world. TBL may be defined as the array of hardware and software used in the teaching and learning systems that include computer-based training systems, multimedia systems, electronic performance support systems, telecommunications systems, as well as the Internet with World Wide Web systems. The rate at which the Internet is being accessed keeps increasing at lightning speed. TBL can enhance teaching and learning; it has the potential to become cost-effective as it offers greater flexibility regarding time and location of training delivery (Furst-Bowe, 1996). Additionally, TBL may facilitate institutional policy regarding access and equity (Lafreniere, 1997). Technology also provides greater flexibility to adapt teaching and learning to meet learners' cognitive and learning styles.

Although ICTs are by far the most significant element undergirding the foundation of TVET, there is a paucity of literature and research regarding its implementation and use in this field of education and training. Attwell (1999) noted: "whilst there is a wealth of studies and debate on the use of information and communication technologies in university and higher education, there has been only limited work on the potential impact for vocational education and training". Even the database of UNESCO-UNEVOC International Centre for Technical and Vocational Education contained very and research for this section, every attempt was made to locate publications related to TVET. However, publications focused on general education were reviewed when no suitable TVET specific publications could be located. The reader can make appropriate inferences regarding the relevance of the discussion to TVET.

Imel (1998) identified four different application of ICTs in adult education, namely: technology as curriculum, technology as a delivery mechanism, technology as a complement to instruction, and technology as an instructional tool. Following is a brief description of each approach.

TECHNOLOGY AS CURRICULUM

When using technology as curriculum the focus is on developing ICT literacy skills. There are two types of ICT literacy skill sets. The first is generic ICT literacy skills such as keyboarding, word-processing, using databases, using spreadsheets, desktop publishing and using the Internet for research and communication (Kasworm and Londoner, 2000). In this network economy every graduate from TVET programmes needs to possess these essential and generic ICT literacy skills. The second ICT skill sets are the occupationally specific ICT literacy skills. Examples of these skills include the ability to use CNC equipment, work with CAD/CAM, and operate equipment with digital system controls.

Technology as Delivery Mechanism

When technology is used as a delivery mechanism the focus is on packaging course content for digital delivery. Common approaches in current use include: computer-assisted instruction (CAI), computer-based instruction (CBI), and web-based or online instruction. Open and Distance Learning programmes make extensive use of technology as their delivery mechanism.

Technology as a Complement to Instruction

When technology is used to complement instruction the emphasis is on providing opportunities to practice skills taught and extending learning by working with specific software applications (Kasworm and Londoner, 2000). Simulators are often used in TVET to address safety concerns during the initial phase of training and to offset cost in renting equipment for training crane operators and truck drivers. In its simplest form, technology can be used for drill and practice to complement instruction.

Technology as an Instructional Tool

Human learning is a very complex process. In spite of years of research in education, our understanding of how human learn is still limited. For this reason educators strive to use the little that is known about human learning whenever they engage in the act of teaching and learning. The learning process can be divided into two broad categories. The first relates to learning conditions that are internal to the learners. While this is the area where the potential to improve learning outcomes is the highest, it is undoubtedly the area that is most difficult to affect. Refer to section 6.3 on Cognitive Divide. The second condition of learning is external to the learners. People learn through the five senses and the contribution of each to the amount that we learn varies.

Following is estimated amount of learning from the five senses (Kupsh and Mason, 1986):

- Taste: 1%
- Touch: 1.5%
- Smell: 3.5%
- Hearing: 11%
- Seeing: 83%.

COMMON TYPES ICTS FOR TEACHING AND LEARNING

The evolution and development of ICTs has resulted in a paradigm shift in the educational system. ICTs are changing the way people learn, offering new alternatives to the traditional classroom. In this new economy, it is essential for learners to have access to education anytime and anywhere. Haddad and Draxler (2002) state, that "lifelong learning and training for the workplace cannot be confined to the traditional classroom. It is unrealistic and unaffordable to continue to ask learners to come to a designated place every time they have to engage in learning". ICTs allow the delivery of education to adapt to an individual needs as opposed to having the individual adapt to how the education is delivered. Education will not be confined to the four walls of a classroom but to wherever and whenever the learner deems appropriate. "Education will not be a location anymore, but an activity: a teaching/learning activity" (Haddad and Draxler, 2002,). ICTs have the capability of providing "personalized, just-in-time, up-to-date, and user-centred education activities" (Haddad and Draxler, 2002,).

Besides their potential to providing education to anyone, anytime and anywhere, ICTs have encouraged new research and development in teaching and learning techniques. As stated by Dean (2002), Schank, the founder of Cognitive Arts, believes that educational institutions must adopt a new way of teaching. He claims that students "learn better through experiential and emotional learning rather than through memorizing names and dates" and thus educators must simulate real-world environments (Dean, 2000,). Technology can be used to facilitate these types of environments. An example of this is using flight simulators to train novice pilots (Dean, 2002). On the other hand, Aldrich argues that this type of learning is restrictive and instead believes in "the ant on the basketball model" where learners have the freedom "go wherever they want but stay in a finite world". Courses would be customized to each user and provide a wide array of options, "like a computer game" (Dean, 2002,). Although these two methods offer two conflicting views of learning, the fact that this discussion is taking place and experiments are being done to determine new methods of learning has intrinsic benefit of its own; experimentation will eventually lead to new and innovative methods being developed.

There are a variety of different technologies that can be used in TVET. Each of these technologies has its own redeeming qualities and limitations, and different situations call for different technologies.

AUDIO-CASSETTE TAPES

Audiotapes can be played by any standard cassette player. Through audio, the tapes can convey information that may be easier to illustrate with sound than simply through text or diagrams. It can accompany other means of instruction (print-based material, classroom teacher, etc.) and provide detailed information step-by-step (Perraton et al., 2002).

Advantages and Disadvantages

Tapes are recordable and thus can be reproduced easily and cheaply. Tapes provide a permanent and flexible storage of information. The tapes can be stopped to allow for classroom interaction and discussion and they can be replayed in instances where it might be beneficial to repeat parts of lessons. Radio programmes can be recorded to allow the lessons to be replayed at a later date (Perraton et al., 2002). Since tapes are relatively small they are portable and can be transported to different areas through the postal service or other delivery service.

A development team is required to develop the lesson plans, edit and record the tapes, and integrate the tapes appropriately with accompanying print-based material or other media. Audiotapes provide only one-way communication where an external instructor cannot interact directly with students and they cannot gauge the progress of the students. The lessons on the tape must also be interactive. To increase effectiveness, the lessons must contain pauses that allow students to think and discuss. Tapes lose their effectiveness if the information is presented as one long lecture (Perraton et al., 2002). It is difficult to present complex concepts using tapes (Nunes and Gaible, 2002). Another limitation with tapes is that the sound quality depends on the quality of cassette player and the "age" of the tape. The sound quality on tapes degrades after repeated use and recording (Perraton et al., 2002).

RADIO

Radio is one of the oldest technologies used for distance education (Stevens, 2001). Radio programmes can be broadcast or interactive (Oujo, 1999). Broadcast radio mirrors the traditional classroom-based model where an instructor lectures through the radio programme and students typically follow with print materials (Oujo, 1999). It can be thought of as "strict" one-way communication where students are not expected to respond and therefore it is hard to gauge the progress of the students (Oujo, 1999). Interactive radio instruction (IRI) can be described as an interactive lesson where an external teaching element is involved in classroom activities via radio (Dock et al., 1999). IRI allows the students to participate as the lesson progresses. To be interactive, a lesson can have spaces or pauses where students can think, develop responses, discuss with other students, or have time to let information sink in.

Radio instruction involves the production, transmission, and reception of the radio programmes (Dock et al., 1999). There are different types of radios available to learners: electric radios (e.g. transistor radios), battery-powered radios, and solar-powered crank radios (Oujo, 1999). Thus, students and the educational organizations can choose a radio that best suits their situation based on the availability of electricity and the supply of batteries. In some cases it may be more cost-effective in the long term to use a solar-powered crank radio. A local or national broadcaster is required to transmit the programme at a particular frequency and specific time (Perraton et al., 2002). The production and development of a programme involves planning, scriptwriting, radio production, and piloting of the programmes by a development team (Dock et al., 1999).

Advantages and Disadvantages

Radio is advantageous since it is relatively inexpensive for learners in developing countries and it is accessible for the illiterate and the poor (Oujo, 1999). Due to the availability of inexpensive radio receivers, it has the potential to reach a large number of learners (Stevens, 2001).

Although radio instruction is considered to be interactive, it is still a one-way communication medium from the external teacher to the learner. The interaction is with printed materials, other students, and the classroom teacher. The material presented in the lesson is only available at the time of the broadcast unless it is recorded. Some other limitations of radio instruction include the inconvenience due to the rigidity of programme schedules and the dependence on "regular and reliable transmission facilities and broadcasting infrastructure" (Perraton et al., 2002). As with cassette tapes, radio is audio-based and thus not always the most effective means for presenting complex concepts.

VIDEOTAPES

Videotapes can be produced and edited by using a portable video camera and a home video editing suite. The technology for video production continues to improve with the emergence of digital video cameras and new computer software that can facilitate the editing process. Videotapes can be viewed by using a TV and VCR combination. It is recommended that videotapes be no longer than 10 minutes with short segments of less than three minutes (Hampton, 2002).

Advantages and Disadvantages

Videotapes appeal to both audio and visual senses. Real-life situations can best be presented and described using video as opposed to using text or audio (Nunes and Gailbe, 2002). Videos can be used when "introducing a new theme to motivate and contextualize learning, after a topic has been addressed in a few class periods to aid students in applying the knowledge they acquired, or after an entire module is completed to show connections to other subjects and disciplines" (Nunes and Gailbe, 2002,). The learner has the flexibility to replay, pause, and rewind videotapes and can repeat lessons as often as they wish (Hampton, 2002). Videotape can support the teaching of practical skills delivered through open and distance learning. Other benefits include "ease of use for the learner, low duplication costs, relatively wide access to the playback technology, and educational effectiveness for imparting practical information" (Stevens, 2001,). Another advantage is that captions and visual cues can be used in the videotapes to help the hearing-impaired.

Caution must be exercised when developing videotapes so as not to include too much information as that could degrade the learning experience. Practical skill must be taught in a clear, concise manner. This strays from the typical teaching model where instructors are taught to provide plenty of explanations and examples. These extra explanations and examples are better served in a face-to-face environment rather than on videotape.

CD-ROM AND DVD

CD-ROMs (Compact Disc-Read Only Memory) store information digitally and they can be used on any computer equipped with a CD-ROM drive (Hampton and Bartram, 2002). DVDs (Digital Video Disk or Digital Versatile Disk) are similar to CD-ROMs and can be used the same way as CD-ROMs but contain more information.

Most CD-ROMs have 650 or 700 megabytes storage space whereas most DVDs have room for 4.7 gigabytes, which equals approximately seven times more storage space than a CD-ROM. DVDs are not widely used yet, mainly because of different standards for writing to DVDs. Once the problem of conflicting standards is resolved, DVDs will probably supplant CD-ROMs as the preferred portable storage medium.

Advantages and Disadvantages

CD-ROMs have a large capacity and can support the storage of information in a variety of formats including text, animation, video, audio, and graphics. Thus, learning materials can be presented in different ways. This allows the material to cater to multiple styles of learning (Hampton, 2002).

For example, Hampton (2002) states:

Learners might first view a video, which provides an overview both visually and audibly of the practical skill they are acquiring. The video could be followed by clear digital photographs of the step-by-step procedure with either text prompts or a voice-over to provide additional explanation. An overview of the complete process could occur in an animation, and then the whole process summarised in dot points using text.

Since the material is stored digitally on CD-ROM or DVD, it is very durable and the quality does not degrade after repeated use. However, scratching the surface or other abuse of the medium may prohibit it from being read by the CD-ROM drive.

A major limitation with CD-ROM and DVD is that a computer with CD-ROM drive (in the case of DVDs, a DVD-drive) is required to access the information. This equipment may not be available to learners in developing countries. Depending on the complexity of the instructional content, a team of experts may be required to develop high quality product. The development team could include a computer programmemer, graphic artist, content specialist, and instructional designer.

INTERNET/WEB-BASED TRAINING

Internet/Web-Based training provides an environment where students access and study course materials online. It may involve the use of live e-learning tools such as application sharing, Internet telephony, online whiteboards, break-away rooms, discussion boards, and chat and messaging programmes that allow real-time interaction between instructors and learners. It can also be used to transmit text, graphics, images, animation, or video. The required tools for online learning include a personal computer and an Internet connection. There are several ways a user can connect to the Internet: standard analog modem (for example, 56 Kbps), Digital Subscriber Line (DSL), cable modem, Integrated Services Digital Network (ISDN), Local Area Network (LAN), cellular, and wireless broadband (fixed wireless and satellite). All connections except for a standard analog modem connection are considered broadband connections. All of these methods allow connection to an Internet Service Provider (ISP) that provides a gateway to the rest of the Internet. An analog modem and ISDN required a "dial up" connection where a user must dial in to connect to the ISP, whereas the other Internet access methods, denoted as "always on" connections, required no dialling.

Wireline Technologies

Analog Modem: This is the original device that allowed computers to connect to the Internet. Analog modems can be internal to the computer or an external device that connects through a serial port on the computer. The standard modems have the capability of transmitting at 56 Kbps. Since this relies on older transmission technology, analog modems provide the slowest connection speed of all other Internet access methods. However, it is relatively inexpensive and it is still widely available to most people. (Comparing..., 2001).

ISDN: Similar to DSL, ISDN uses a normal copper telephone line to connect to the Internet. The copper line must be "converted" to a digital ISDN line, after which one can connect to the Internet and have a phone conversation over the same line. Basic service enables connection speeds up to 128 kbps. An ISDN modem is used to interface a computer to the telephone network. The modem is either an internal card or an external device that connects to a computer's serial or USB (Universal Serial Bus) port (ISDN, 2000) (broadband 101 - ISDN..., 2001). ISDN is also a "dial up" service and users must dial in to their ISP (broadband 101- ISDN..., 2001).

DSL: DSL uses a normal copper telephone line (a pair of copper wires) to provide a high-speed connection to the Internet. Thus, it is telephone companies that typically provide DSL connections. When connected to the Internet, the phone line can still be used for normal voice calls. A DSL transceiver (modem) is the interface between the computer and the phone line. The computer is connected to the modem either through a USB or 10 base-T Ethernet. If using an Ethernet connection, the computer must have an NIC installed where the NIC connects to the modem through an Ethernet cable. If the computer has a USB port, it can connect to the modem via a USB cable. There are two different categories of DSL: asymmetric and symmetric. For asymmetric DSL, the downstream speed is faster than the upstream speed whereas for symmetric DSL, the downstream and upstream speeds are the same. Connection speeds vary from country to country and they also depend on the particular DSL provider within that country. In OECD countries, downstream (Internet to customer) connection speeds range from 128 kbps to 8 Mbps (where speeds of 256 kbps and 512 kbps are the most common) and upstream (customer to Internet) connection speeds range from 64 to 740 kbps (where a speed of 128 kbps is the most common) (Paltridge, 2001).

Cable: A cable connection uses the same coaxial cable used for television signals to send and receive data to and from the Internet. A cable modem acts as the interface between the computer and the coaxial cable. The computer is connected to the modem either through USB or 10 base-T Ethernet. If using an Ethernet connection, the computer must have a NIC installed where the NIC connects to the modem through an Ethernet cable. If the computer has a USB port, it can connect to the modem via a USB cable (broadband 101 – cable, 2001). As with DSL, connection speeds depend on the cable provider within the particular country. In OECD countries, the advertised downstream connection speeds range from 128 kbps to 2 Mbps (where speeds of 256 kbps and 512 kbps are the most common) and the advertised upstream connection speeds range from 64 to 768 kbps (where a speed of 128 kbps is the most common) (Paltridge, 2001). *LAN*: The computers in businesses and educational institutions are usually connected to a LAN. A LAN connects computers and peripheral devices using the same transmission line, allowing for applications, devices (for example, printers), and data storage to be shared among the members of the LAN. There are three different methods for connecting computers in a LAN: cables, wireless, and power line systems (Rusten and Hudson, 2002). Cabled LANs use cables and hubs to physically connect the computers and devices. A hub controls the flow of network traffic between individual computers and the system's server, usually in clusters of 20 to 30 cables. Wireless LANs do not use cables but rather a wireless network adapter (receivers) that is either an internal network card or an external device that connects to the computer through USB. As well, "one or more wireless network hubs/transmitters are connected to the server" (Rusten and Hudson, 2002, p. 86). Power line LANs use the existing power lines in an institution for the transmission of network traffic (Rusten and Hudson, 2002).

LANs are typically connected to an ISP through a high-speed T1 line provided by a local telephone company. A T1 line is either a fibre optic cable or a copper line and can carry a data rate up to 1.544 Mbps. All the users that connect through the same T1 line share this bandwidth. Larger companies need higher capacity lines such as T3, OC3, OC12, OC48, and OC192. The higher capacity lines run on fibre optic cables.

Optical fibre: Optical fibres provide a huge amount of bandwidth that is ideal for high-performance applications such as videoconferencing or web-based applications that rely on a large amount of graphics (Rusten and Hudson, 2002). As mentioned previously, telephone companies provide these high capacity lines. They are typically installed for "institutional customers such as hospitals, schools, and businesses" (Rusten and Hudson, 2002,).

Wireless Technology

Cellular: A cell phone with a cellular modem can be used to connect to the Internet. In some developing countries (such as Cote d'Ivoire, Gabon, Rwanda, Tanzania, Uganda, Cambodia, and the Philippines), cellular telephones are more widely used than wireline phones. However, cellular access can be expensive and the bandwidth is limited. Thus, it is more suitable for "short bursts of use" such as e-mail than running web-based applications (Rusten and Hudson, 2002).

Wireless Broadband: There are two types of wireless broadband service: fixed wireless and satellite (Wireless, 2003).

Fixed Wireless – A fixed wireless broadband connection uses a ground-based antenna system. A small rooftop antenna communicates with transmission towers, which are directly connected to the Internet (Keizer, 2001). A modem or receiver connects to the exterior antenna via coaxial cable and also connects to a NIC installed on the computer through an Ethernet cable (Wireless, 2003). (Broadband 101 – Wireless..., 2001). Downstream connection speeds can reach 5 Mbps but are usually around 1 – 2 Mbps. Upstream speeds are around 256 kbps (Keizer, 2001).

Satellite – For high-speed satellite Internet connections, a satellite dish and a "satellite modem" are required. Data signals are sent from the satellite dish to satellites orbiting the Earth. The dish is about 2-foot by 3-foot in size and the modem acts as the interface between the satellite dish and the computer. A coaxial cable connects the dish to the modem (How does..., 2001). The latest satellite technology allows for downstream and upstream speeds up to 400 kbps. For older satellite service, one may also require an analog modem and another ISP for upstream data transfer. Using an analog modem degrades upstream rates to the speed capability of the modem. As mentioned previously, analog modems have a maximum data transfer rate of 56 kbps.

ICTS FIT IN THE LEARNING ENVIRONMENT

Haddad and Drexler (2002) state that there are five different hierarchical levels of education where ICTs can be used: presentation, demonstration, drill and practice, interaction, and collaboration.

OPEN AND DISTANCE LEARNING IN TVET

UNESCO (2002a) defines Open and Distance Learning (ODL) as an "educational process in which all or most of the teaching is conducted by someone removed in space and/or time from the learner, with the effect that all or most of the communication between teachers and learners is through an artificial medium, either electronic or print" (p. 22).

IITE (2000a) identified three types of ODL:

1. Learner-content;
2. Learner-learner; and
3. Learner-tutor.

The infrastructure and level of technology available in the region determine the choice of the medium. This may include printed study guides, educational television, radio systems, multimedia systems, and Internet-based systems. The open nature of ODL helps learners take responsibility for "what they learn, where they learn, (and) how quickly they learn".

Mishra (2002) noted that "advances in education technology have led us to accept that the benefits from TVET through ODL are far greater than from other types of courses". Hampton and Bartram (2002) argued:

If access to technical and vocational education and training (TVET) is to increase, new ways of developing and delivering courses must be explored. TVET must be taken outside of the classroom and into the communities, the workplaces, and the homes of the students. Traditional ways of thinking about TVET must be put aside and different ways of packaging and delivering knowledge and skills must be developed. This is particularly critical in meeting the demands of countries with depressed economies and countries where people are separated by water (as in island states) by terrain or by distance…

ODL can be used in TVET to empower those people most disadvantaged such as the disabled, women, and the unemployed and ethnic minorities. ODL can also allow greater participation in technical and vocational education and training after work hours for those who can't afford to take time off from their jobs and who are interested in improving their vertical mobility professionally (Mishra, 2002). ODL in the field of vocational training may have to be supplemented by hands-on work though, which could be done through residential schools, home experiment kits or collaboration with the workplace (UNESCO, 2002a).

John (2002) noted that in developing countries ODL has been successfully used to change the social lives of people, raising them from economic vulnerability to economic empowerment. ODL has been used to successfully impart carpentry skills, building construction, etc. Dhanarajan (2002) also noted that ODL has been successfully used for training workers in many fields ranging from farming to electronics, health to engineering, and animal husbandry to automobile engineering. There are many obstacles that hinder the full implementation of distance learning. Some of these barriers include: lack of infrastructure, underfunded programmes, high costs of installation and maintenance, lack of organizational support, and lack of training of those involved.

Some factors affecting costs of ODL include: number of course materials that need to be developed, the frequency with which course material must be revised and changed, and the choice of technology. Advocates of ODL often cite economies of scale in their support of this mode of delivery. However, this concept does not necessarily apply to distance learning. There are many cases where it would be more cost efficient to attend classes on campus instead of studying through distance education. Many North American universities that invested heavily in distance education are now phasing out their programmes due to low enrolment and a high dropout rate. It appears that the role of distance education is not to minimize the costs of teaching and learning, but to find various media able to reach as many people as possible (UNESCO, 2002a).

ICT-MEDIATED LEARNING ENVIRONMENTS

ICT-mediated learning provides two educationally rich learning environments to the learners:

1. Explorative environments; and
2. Interaction environment (Dillemans et al. 1998).

In the explorative learning environment learners can practice their skills in safe settings, but they can also investigate (simulated) aspects of the world. Modelling tools go even further, enabling students to shape (simulated) the world themselves (Van Baalen and Moratis, 2001). In the interaction environment ICTs facilitate human interactions. Dillemans (et al., ibid) proposed six critical dimensions that must be examined in order to bring out the best of educational technology in the explorative and interaction environments.

ICTS THE NEW ECONOMY: CHALLENGES AND OPPORTUNITIES

ICT is the driver of the new economy. Therefore an analysis of the use of ICTs in TVET would be incomplete without a discussion on the new economy. The purpose of this section is to demystify the new economy. It is based on the recent work of Chinien, Moratis, Boutin, and Van Baalen (2002).

In recent years, the term "new economy" has increasingly been accompanied by negative connotations. Whether they see it as "an Internet bubble that is about to burst" (or that has already burst) or blame it for "the worldwide collapse of economies", adversaries have tried to demonstrate the negative effects and even the non-existence of the new economy. It is undeniably true that discussions on the new economy have in many aspects been dominated by near adulation, often by business gurus and fad-chasing politicians. The new economy, they claim, is a true and unprecedented economic revolution.

Meanwhile, different authors have posed the questions "What is new about this economy?" and "What is this new economy about, really?" (For example, see Webber, 1993) Now that the limits of the economy seem to have been reached, it is time to realistically reflect on what this "new economy" has brought and, indeed, what it is about.

DEFINING WHAT THIS NEW ECONOMY IS

Several conceptions of what has become widely known as the new economy have surfaced over the years. Many of these conceptions seem to acknowledge that this new economy refers to an economic era that is characterized by a new set of economic activities, economic volatility, new organizational architectures and paradoxes. It has engendered economic industriousness of a totally different nature and of distinct dimensions. What makes it not so radically new, however, is that, according to the theory of Kondratieff cycles, such a transition takes place about every 40 to 60 years. This new economy is not the first new economy.

Agricultural activity dominated wealth creation for about eight thousand years before largely coming to its end with the first industrial revolution, which took place at around the end of the eighteenth and the early nineteenth century. The new economy may be represented as a comparable economic transition driven particularly by revolutions in technology. As illustrated by Norton (1999), global development unfolds through the succession of new economies, underpinned by the Schumpeterian process of creative destruction. The new economy that people have been talking about over the past few years is more a reborn economy than a totally new economy, meaning it has successfully weathered what could be termed as a maturity crisis and has challenged relatively recent predictions of economic decline (Norton, 1999).

Two global trends seem to particularly explain this new economy: globalization and the widespread use of ICTs. Globalization has led to the proliferation of market forces, freer trade, and widespread deregulation, while ICTs are creating new companies and new industries before our eyes (Shepard, 1997). Castells speaks of the prevalence of intertwined global and informational features. They are informational because productivity and competitiveness of units or agents fundamentally depend upon their capacity to generate, process, and apply knowledge-based information; they are global because the core activities of production, consumption, and circulation are organized on a global scale. But what structural features are these drivers of this new economy bringing with them? A good starting point for exploring the more structural features of the current new economy is found in Wired magazine's Encyclopaedia of the New Economy:

When we talk about the new economy, we're talking about a world in which people work with their brains instead of their hands. A world in which communications technologies create global competition – not just for running shoes and laptop computers, but also for bank loans and other services that can't be packed into a crate and shipped. A world in which innovation is more important than mass production. A world in which investment buys new concepts or the means to create them, rather than new machines. A world in which rapid change is a constant. A world at least as different from what came before it as the industrial age was from its agricultural predecessor. A world so different its emergence can only be described as a revolution.

In light of this encompassing definition, this latest new economy can be defined as a knowledge and idea-based economy where the keys to job creation and higher standards of living are innovative ideas and technology embedded in services and manufactured products; it is an economy where risk, uncertainty, and constant change are not the exception but the main rule (adapted from The New Economy Index web site).

STRUCTURAL ECONOMIC FEATURES OF THE NEW ECONOMY

Weinstein (1997) identified some structural features of this new economy in the form of macroeconomic traits, contending that the new economy is an economy that grows without apparent threats of recession; that continues to expand without an increase in inflation; that is constantly restructuring itself for greater efficiency and productivity; that replenishes and revitalizes itself through new technology and capital investment; that functions without excessive debt, either public or private; that maintains a balanced budget; and that is increasingly globalized and export-driven. These attributes are the fundamentals on which a prospering economy can be built. The New Economy Index mentions four themes (industrial and occupational change, globalization, dynamism and competition, and the information technology revolution) encompassing 13 indicators that illustrate the emergence of these structural roots. (The following elaboration is based on the New Economy Index and was adapted from Baalen and Moratis, 2001).

Industrial and Occupational Change

Within this first theme, several trends can be witnessed. First of all, more people work in offices and provide services. Decreasing amounts of time are spent on physical labour, whereas knowledge-intensive work, requiring analytical and interpretive skills, is increasing. This is linked with the rising importance of the service sector in Western economies. For instance, in the US 80% of the workforce do not spend their days making things; they work in jobs that require them to move things, process or generate information, or provide services to people. Second, high-wage and high-skill jobs have grown, as have low-wage and low-skill jobs. Knowledge-based jobs (requiring post-secondary, vocational or higher education) are growing as a share of total employment both in the US and Europe. In the US, managerial and professional jobs increased as a share of total employment from 22% in 1979 to 28.4% in 1995.

Globalization

With globalization, trade assumes an increasing share of the new economy, which means more intense competitive structures. As a consequence, firms have to innovate at a constant rate in order to be successful. Innovation means "coming up with new ideas about how to do things better or faster. It is about making a product or offering a service that no one had thought of before. It is about putting new ideas to work in our business and industries and having a skilled work force that can use those new ideas." (Government of Canada, 2002,). The value of the world economy that is "globally contestable" – meaning open to global competition – will rise from approximately 4 trillion dollars in 1995 (which equals about one-seventh of the world's output) to over 21 trillion dollars by the year 2000 (half of the world's output) (Frazer and Oppenheim, 1997). A usually revealing indicator of globalizing tendencies is foreign direct investments (FDI). FDI is an essential condition for business if it is not to become outperformed by future competition in this globalizing world.

Dynamism and Competition

The theme of dynamism and competition poses six trends in the observation of a new economy. First, there is the spawning of new, fast-growing entrepreneurial companies, which indicates the degree of the economy's innovative capacity. The so-called "gazelles", or companies with a sales growth of at least 20% per year for four straight years, are an important indicator of this trend. In addition, investors are increasingly embracing entrepreneurial dynamism by, especially in the sector of information and communications technologies. Second, there is a general increase in and intensification of competition. For example, in 1965, IBM faced 2,500 competitors for all its markets.

By 1992, it faced 50,000. Stable industries have become dynamic and global, while at the same time numerous mergers and acquisitions have taken place, including mergers between corporate giants and acquisitions of the most innovative small- and medium-sized enterprises. Third, just as merging and acquiring are ways of coordinating corporate activity, cooperation among competitors has become ubiquitous. The dynamics of organizing through network structures and "co-opetition" have been labeled the prominent organizing principle in the new economy (for example, see Drucker, 1999). Firms are looking for opportunities to establish partnerships with suppliers, consumers, and universities to enhance product and service development and technological innovation. Fourth, the new economy is constantly churning, indicating the balance between job creation and job destruction (this is called "flux").

Though this constant churning confronts workers with insecurity, instability, and thus economic risk, it is also a major driver of innovation and growth. Fifth, consumer choices are exploding and consumer needs are more specifically addressed. Due to the fact that mass customization and flexible production have replaced mass production and standard products ("one-size-fits-all"), customers now can choose, and order at a fraction of previous prices tailor-made products and services ("one-size-fits-(n)one"). Variety and diversity at low cost are conjoined in the new economy. Sixth, and last within the theme of dynamism and competition, speed is becoming the standard. Competition-induced time-to-market considerations and short product life cycles are characteristic of the new economic order. A well-known example is Moore's Law: every 18 months, the processing speed of microchips is doubled, while prices are cut by half. Business has to run just to stay in place in an economy where people talk about technological development in "web years", which is a quarter of a normal year. Greater consumer choice and the sophisticated nature of interaction between business and customers are just two examples of the possibilities the increase in speed brings.

The Information Technology Revolution

This fourth and final theme concerning the rise of the new economy entails three trends. First, there is the omnipresence of microchips, which is the foremost characteristic of the digital era. In 1984, worldwide shipments of semiconductors totalled 88 billion units, and by 1997 world shipments had nearly tripled. By 2003, the number is expected to pass the 400 billion unit mark. Second, the costs of computing are dramatically decreasing. Computing costs are dropping nearly 25% per year. Since information technology is expected to increase efficiencies, enabling cost cutting, customizing products and services, and increasing the speed of commerce, all sectors of business are investing in ICTs. Therefore, this trend is important as it enables the transformation of business and even whole industries. The third trend co-depends on the previous trend: data costs are plummeting. Global communication is, undoubtedly, one of the major enablers of the economy.

The impacts of this information technology revolution and the fact that ICTs can be seen as one of the major driving forces of the new economy have caused authors to coin the term "digital economy". This term arose from the observation that the relatively smooth transition from the old economy to the current new economy has been caused by the emergence of information goods, which can be digitised. The prominence of information commodities entails three landmark events: the invention of the microprocessor in 1971, the introduction of the IBM personal computer (PC) in 1981 and the commercialization of the Internet in 1994 (Norton, 1999).

The microprocessor switched the world from an analog to a digital mode in which virtually every person, company, and government is a customer for technology products, mostly because of the introduction of PCs in 1981. The definition of technology industries has expanded from large computers to include personal computers, software, semiconductors, semiconductor equipment, communications (both telecommunications and data communications), and medical technology (biotechnology and medical devices) (Murphy, 1998). The coming of the PC thus rendered anything and everything subject to the power of the computer, while retaining the crucial dimensions of human scale, decentralized decision making, customized design, and creativity.

Climate for Growth

Weinstein (1997, cited by Baalen Van & Moratis, 2001) examined the new economy form a microeconomic perspective.

He identified some conditions that were essential for growth in the new economy, namely, an economy:

- That grows without apparent threats of recession;
- That continues to expand without a pickup in inflation;
- That functions without excessive debt; and
- Maintains a balanced budget.

Knowledge and Brain Power

Technology is said to be the driver of this new economy and human capital is its fuel (Moe and Boldget, 2000). In the new economy, human capital is defined as workers' knowledge that results in effective and efficient performance. Knowledge is not only beneficial to the well-being of the worker, but is also viewed as a major competitive advantage for the company and a key element to ensure the country's national prosperity and social development. The perception of the role of human intervention in the economic transactions has also changed. Keursten and Kessels (2002) noted that the "focus is shifting from appreciation of physical labour and the ability to coordinate and regulate to the ability to contribute to knowledge generation and application". This recognition of the importance of knowledge and brainpower for economic prosperity and competitiveness has drawn considerable attention to the following concepts: lifelong learning, self-paced learning, on-site education, just-in-time education – any place and any pace, innovation, and learning to learn.

In Capital and Knowledge as a Wealth Creator

These structural or macro economic roots outline what the new economy seems to comprise and how value is created in this new economy. Technology plays a central role, but the one thing in particular everyone seems to agree on is the belief that this era will differ to the extent to which it will be dominated by knowledge and skills. In fact, the ICT revolution makes knowledge a competitive resource. In this economic era, economic prosperity depends on brains rather than brawn and value creation is realized by means of employing knowledge workers and continuous learning.

In the nineteenth century, capital equipment was seen as the single decisive factor in driving economic growth; knowledge and skills only played a supporting role. Thurow (1999) notes that Adam Smith barely mentions education in his Wealth of Nations and when he does mention it, he disqualifies it as an antidote to the mind-numbing boredom of factory work. Thurow argues that in an economy dominated by knowledge-based activities, two interlocking but differing sets of skills related to the creation and the deployment of knowledge are required. Knowledge creation requires highly educated creative skills at the very top of the skill distribution. Knowledge deployment requires widespread high-quality skills and education in the middle and bottom of the skill distribution. Burton-Jones (1999) comments:

Future wealth and power will be derived mainly from intangible, intellectual resources. This transformation from a world largely dominated by physical resources, to a world dominated by knowledge, implies a shift in the locus of economic power as profound as that, which occurred at the time of the Industrial Revolution. We are in the early stages of a "Knowledge Revolution", the initial impact of which is becoming apparent in the volatility of markets, uncertainty over future direction within governments In the "former new economies", distinctive competencies have included access to raw materials, natural resources, and fertile lands. In an era in which knowledge is the important intangible asset, a national competitive advantage will increasingly be derived from an educated and empowered workforce able to adapt quickly to the dynamic requirements of a changing world (Ives, 1992). From this point of view, knowledge, learning, and education become ever important and move centre stage in developing a competitive advantage, not only at the level of the nation-state, but increasingly at the firm level and, of course, the personal level. Knowledge workers will be valued for their creativity, ideas, flexibility, and adaptability, which requires intelligent and thoughtful human resource management.

This is backed by Robert D. Atkinson, director of the Technology, Innovation and the New Economy Project at the Progressive Policy Institute (PPI), who proposes three main foundations to sustain the growth of the new economy: speeding the transformation to a digital economy, investing in research and innovation, and improving skills and knowledge. The number of engineers, he notes, has grown, but not at the rate the new economy requires. Moreover, intensified competition, churning of firms, and low employee tenure has led companies to spend less on skill training. PPI therefore proposes that companies collaborate in networks, partnerships, and consortia to invest in training, to ensure innovation and growth (Atkinson and Court, 2000).

Effects of Globalization and ICTs on Skills

Contrary to popular belief, the decline of industrial jobs in advanced economies is only partially related to globalization; technological change has had more profound impact on the labour market: "today's workplaces are often in multi locations characterized by cultural diversity" (Lynch, 2002,). Furthermore, in many organizations, relationships among competitors, suppliers and customers "are complex and boundaries have blurred between organizations and their environments" (Hiniker and Putnam, 2002,). Advances in technology have helped to de-massify production, reduce economies of scale and boost industrial productivity to the extent of de-industrializing the economies, thus causing a movement of labour from industry to the service sector. There is a public perception that all service sector jobs are low-skill and low-paid. In reality however, many good and well-paid jobs for highly skilled workers have also been generated by the service sector, especially in more advanced countries (Camdessus, 1997). Therefore, Toffler and Toffler (1995) argued that the current trend for denigrating the service sector as "hamburger flipping" must stop.

In their attempts to become and remain more competitive in this new world order brought about by globalization and technological developments, many enterprises have adopted different approaches to human resource management and organizational practices, such as the following: hiring and layoffs, altering hours of work, using part-time and contract workers, subcontracting and outsourcing work, reducing hierarchical structures, adopting more fluid job designs, as well as implementing multi-tasking, multi-skilling, self-managed teams and multi-functional teams (Human Resources Development Canada, 1996).

There seems to be no consensus regarding the impact of technological change on the workplace and its effects on jobs and skills. Some believe that technology is de-skilling jobs, while others argue that technology is up-skilling jobs. Tenets of up-skilling would argue that in this emerging knowledge-intensive economy the proletariat is being replaced by a cognetariat where work is increasingly shifting away from manual tasks toward symbolic processing (Toffler and Toffler, 1995). The concept of "mind workers" or knowledge workers has emerged from this transformation of work and it appears to be the fastest growing type of worker.

The country studies cited by the International Labor Organization (2001, a) indicate that the introduction of ICTs can contribute both to up-skilling and de-skilling of workers. "On the one hand, ICTs can downgrade skills and competence to single-task machine-tending and, on the other, it can upgrade skills and competence to multi-task work relying on greater creativity". The European Union's Livre blanc sur l'éducation et la formation (1995) also indicates that with the advent of new technologies Europe is moving toward a "cognitive society" in which work content will be limited to tasks requiring initiative and flexibility on the part of the worker.

The re-engineered workplace expects workers to perform "more complex tasks, use new technologies, involve less direct supervision of workers, and acquire the use of higher skill and knowledge levels in making decisions and solving problems individually and as members of collaborative work teams" (Frantz, Friedenberg, Gregson and Walter, 1996,). As Johnston and Parker predicted in 1987, in the US there were some indications that 42% of new jobs were going to require high-level reasoning in the new millennium. Lynch (2002) reporting today's figures indicated that "less than 20% of the workforce is in jobs classified as unskilled" as opposed to 60% in 1959; the remaining 80% of the job market are skilled occupations (60%) and professions (20%).

The synergy of combining new technologies with human skills in the post-industrialized society has created a controversy regarding the effects of technology on skills. Those who posit that technology has augmented skill requirements, advocate increased specialized job skill training. On the other hand, proponents of the de-skilling concept argue that many trade skills traditionally imparted in Technical and Vocational Education and Training (TVET) are no longer valued in the high performance workplace. Instead, they recommend the development of solid basic foundation skills as well as generic, broadly transferable employability skills (Budke, 1988). This trend is more apparent in the service sector where entry requirements are less well defined (Quebec, Ministry of Education, 1998). This conflict in views "exacerbates a traditional split among educators and policy-makers concerning the appropriate role and function of job training and vocational education" (Budke, 1988,). There is evidence that ICTs have polarized jobs into highly skilled and unskilled.

Dignity of Work and Fair Pay

For some people, work over the past decade has become technologically complex, demanding sophisticated work skills. Others, as a result of unemployment, have been denied the satisfaction that comes from the challenge of work, and the sense of achievement gained. The job market and jobs have become sharply polarized by a simultaneous increase in jobs requiring post-secondary education and casual jobs for unskilled workers. This polarization of the labour force is widening the inequalities between the haves and the have-nots and between the educated and less educated. Tuijnman (1997) noted: "Alienation, social exclusion, unemployment and slow productivity growth are among the consequences of limited capacity of some countries to adapt and fully exploit the opportunities offered in the global knowledge economy".

The ILO estimated there were 160 million unemployed workers at the end of year 2000; most of them were first time job seekers, with as many as 50 million coming from industrialized countries (ILO, 2001). Moreover, many people are currently forced to accept low-skill, low-paid, part-time and non-permanent jobs. ILO (2001, a) reported that about 500 million workers, mostly from developing countries, do not earn enough to keep their families above the poverty line. Technological innovations, which call for highly skilled workers, are contributing to widening inequalities in income and employment prospects between the skilled and less skilled workers (Camdessus, 1997). Even the developed countries have not been spared.

Maxwell (2000) noted: "Despite all the talk about the explosion of knowledge-based economic activity and its contribution, two million adult Canadians work for less than $10 an hour–about one in six employed people". Surprisingly, many of these people are well educated. One-third has a post-secondary diploma or degree. The US is experiencing the same kind of phenomenon. The results of a number of recent surveys indicated that the rate of underemployment among four-year college graduates in the country range from 40% to as much as 50% in the social sciences. The underemployed are found among the young highly educated workers (Gray, 2002).

Hobart (1997) noted that: "unemployment is one of the most pressing problems facing the world today...every government around the world has to grapple with this issue". This economic dislocation is having a devastating effect on young people who enter the labour market without any marketable skills. Young people are being disproportionately affected by high unemployment, underemployment, and credential inflation. In the OECD countries many young people work in "non-standard" part-time and insecure jobs primarily in hotels and restaurants, wholesale and retail trade, and automotive repair to acquire experience that they hope will lead to better jobs (Quebec, Ministry of Education, 1998).

While traditionally youth unemployment has just been considered "a rite of passage in growing up and making one's way in the world of work" (Canadian Youth Foundation, 1995, a,), it has now reached such a tragic proportion that it has become the number one priority on governments' agenda. According to ILO (2001, a), the youth unemployment rate in OECD member countries in 1999 was 11.8 per cent. In the US, it is estimated that as much as 30% to 50% of students enrolled in pre-baccalaureate technical education across the country are underemployed college graduates or college drop-outs (Gray, 2002). According to Gray (2002) "there is a surplus of individuals with degrees but also a shortage of university graduates with skill". Although the full economic, social, and psychological impacts of chronic youth unemployment are not fully understood, its perceived threats to social stability are drawing considerable attention.

Various initiatives are being implemented to facilitate the transition of young people from school to work. A "school-to-work" plan can be defined as a "system that would equip all young people with high levels of academic and occupational knowledge and skills to find employment that uses their capacities" (Stern, Bailey and Merritt, 1996,). Many successful school-to-work programmes have been implemented in Austria, Canada, Germany, Japan, Switzerland, and the US. Common types of programmes implemented in Canada include mentoring programmes, programmes targeted to special needs groups, career counselling and job placement, school-business partnerships, co-operative education, restructured programmes, youth apprenticeship programmes, and supervised work-based learning. The Norwegian Youth Guarantee programme is perhaps the most comprehensive approach to address youth alienation to the labour market. Under this programme, unemployed youth are offered a job (with or without training) for a six-month period (Quebec Ministry of Education, 1998). Research shows that women and older workers have also been severely affected by the economic change and workplace transformation (Human Resources Development Canada, 1996).

While a sound school-to-work policy is a step in the right direction, by itself it will not solve youth unemployment. All social partners should work together to create economic climates favourable to job creation (Association of Canadian Community Colleges, 1998). The Canadian Youth Foundation study (1995, b) concluded that at the rate at which new jobs are being created in Canada, self-employment remains the only entrance to the world of work for many young Canadians. The Jua Kali programme in Kenya is an excellent example of how self-employment can also alleviate poverty in developing countries (Mbugua, 1997). TVET should play both a reactive and proactive role with regards to unemployment. Reactively, TVET must collaborate with its social partners in helping the unemployed, especially youth and other disadvantaged groups, to acquire the critical academic and occupational skills necessary for employment. Research shows that people with inadequate skills, females, young people from disadvantaged regions, migrants and people with handicaps tend to have a more difficult time during the transition to the post-industrialised society (Reubens, 1981).

Proactively, TVET should focus on the implementation of school-to-work programmes to facilitate the transition of young people to the workplace. An analysis of recent international initiatives identified the four underlying principles of effective school-to-work programmes: new curricula should be created that integrate vocational and academic studies; occupational and educational performance standards should be explicitly related to each other; initial education and training should include a certain amount of work-based learning for all students; employers and educators, including both vocational and academic educators, must share both responsibility and power in new school-to-work systems (Stern, Bailey and Merritt, 1996,).

DIGNITY FOR FLEXIBLE ACCESS TO TVET THROUGHOUT LIFE

Workers of the future may expect to work on short-term assignments, on a contract basis, or within several project teams. They may have to work for more than one employer at any given time or even in new careers several times. Additionally, the Canadian Labour Market and Productivity Centre (1993) estimates that fifty per cent of technical workers' skills become obsolete within three to seven years. Lifelong learning is the only way to prevent obsolescence and remain competitive in a job market where work is becoming increasingly knowledge-intensive (Pritchett, 1995). One of the challenges with regards to lifelong learning is to provide recurrent opportunities for access to education and training throughout the life of a person (IITE, 2001a).

The need for recurrent education and the changing labour market conditions call for flexible access to TVET. Continuing education models that will meet workers' lifelong learning needs have to be relevant and flexible to provide just-in-time learning without distance (Manitoba Education and Training, 1998). In response to this need some institutions are implementing open-entry/open-exit programmes in TVET. For example, UNEVOC (1998) reports that Sweden has a very extensive system of adult education, one that is flexible enough to allow adults to acquire their education at different stages in life through various modes of delivery.

While there is an official discourse about the need for lifelong learning, accessibility still remains a major issue. John Daniel, Assistant Director-General for Education, UNESCO recently noted:

The problem in education today is that hundreds of millions of the world citizens do not receive it. Many more do not get enough of it.... Over 100 million children never see the inside of a school. As many more do not stay in school long enough to gain useful skills. 800 million adults have their lives blighted by illiteracy (2002,).

In the context of TVET, Stevens (2001) pointed out that: "distance education is believed by many to hold promise in addressing critical problems facing skills development at present, namely: a lack of qualified instructors, the need to greatly increase the delivery of skills training on a wide scale, and the need to deliver training at a much lower unit costs owing to constraints on financing".

In commenting on the use of ICTs to increase access to education, Haddad and Draxler (2002) noted:

It is going to be very difficult-if not impossible for countries to meet the objective of effective learning, for all, anywhere, anytime. Our inability to meet this challenge, however, is self-inflicted because we tend to think of linear scaling, that is, using the same model of education (a school constrained by space and time) but more of it and on a larger scale. What we really need is to think differently and radically. The education model developed for the Industrial Age cannot achieve educational empowerment effectively in the Information Age. With ICT tools, we should be able to evolve the components of the conventional model into the corresponding components of the new model. Limited access to TVET particularly in rural areas and for disadvantaged groups continues to be a major issue for most countries in Asia-Pacific region. The following suggestion was made to address this issue at a Regional Planning Meeting of TVET experts (UNESCO-UNEVOC meeting, 2002):

The TVET system should introduce poor-friendly training, delivered at their doorsteps at times convenient to them through appropriate modes of training. It may take the form of mobile training, distance learning, strengthening training by master craftsmen and such other innovative modes of training delivery.

While ICTs hold promises for improving access to education there is also growing concerns regarding the divide that is creating among people. CONFINTEA V (Fifth International Conference on Adult Education) "warned about the emerging form of exclusion created by the development of new ICTs" (UNESCO, 2003). TVET has an important role to play in developing the computer literacy of students. Most vocational students are from disadvantaged working class or minority students having limited access to home computers. The integration of computer-based technology in TVET would provide equality of opportunity for these students. Passmore (1983) notes: Vocational education classrooms are where large proportions of low-income, minority, and female students (home economics and business education programmes typically have high female enrolments) can be found. It therefore seems appropriate to educate these students in the computing skills, which will be necessary in their employment and in the work of the household.

INTERACTIVE TELEVISION

In this context, interactive television refers to instruction occurring over broadcast television. It allows learners to receive live television instruction remotely, away from the actual instructor. The instructor(s) are located at a broadcast studio and the learners view the instructor(s) on a television monitor. Interaction is provided by one or more additional components. "They can ask questions and/or provide feedback to the instructor through a number of mechanisms that can be used either independently or in combination" (Stevens, 2001,). Typically, an audioconferencing mechanism is used for real-time interaction. Additionally, a response pad can be used to link learners to a computerized instructor console in the broadcast studio (Stevens, 2001). There are other asynchronous aids that can also be used such as e-mail and fax. The actual broadcast can be achieved using geo-stationary satellite, microwave, cable, or fibre optics. There are also commercial suppliers that provide the infrastructure and programme development services for interactive television. Learners usually congregate at an interactive television site that has all the required equipment (such as satellite receivers, teleconferencing tools, and television monitors or projection systems) to receive instruction (Stevens, 2001).

Advantages and Disadvantages

The main advantage of interactive television is that instruction can be transmitted to several different sites, and thus potentially reach a large number of learners using existing broadcasting infrastructure. It can provide a high level of interaction if an audioconferencing mechanism is utilized. The effectiveness of the instruction depends on the quality of the audiovisual equipment. Low-quality equipment and unreliable infrastructure that causes outages can seriously hamper instruction (Jurich, 2003).

The main drawback of an interactive television system is its high cost, both at the main broadcasting site and the learner sites. There is also a high telecommunications cost for the transmission of the video, and there may be an additional audioconferencing cost.

These costs, however, can be offset if the instruction can reach a large enough number of people in multiple sites. Thus, travel costs, number of staff and staff hours can be reduced significantly (Stevens, 2001).

VIDEOCONFERENCING

Videoconferencing allows participating individuals in different locations to see and hear each other in real-time through videoconferencing equipment (Stevens, 2001). Dixon (2000) states that it "is much more like using the telephone, with the added feature of being able to see the person you are talking to".

Older videoconferencing technology uses ISDN or leased telephone lines at high hourly rates and the installation cost of this type of system is expensive. It is also a more rigid environment that requires special room set-up, advanced scheduling, and a professional operator (Dixon, 2000).

The more recent videoconferencing technology, however, uses the Internet as the transmission infrastructure, thus decreasing the installation and usage costs. It also has the same advantages as the Internet: flexibility, convenience, and ubiquitousness.

A videoconferencing terminal consists of several components (Videoconferencing..., 2002):

- *A camera*: Captures live movement of the participating parties;
- *A video display*: Displays the images of the other remote parties;
- *A microphone and speakers*: Enables the transmission and reception of voice and audio;
- *A CODEC (CODer/DECoder or COmpressor/DECompressor)*: Digitizes the audio and video information appropriately to allow for efficient, rapid transmission over the network. CODEC can be either a software or hardware component that enables real-time communication;
- *User interface*: Provides the mechanism to install and uninstall the system, place calls, control audio/video features (for example, volume, echo, color, brightness), etc;
- *Computer system*: Can be either the typical desktop PC or a standalone system;
- *Network connection*: The new video conferencing standards (H 323) were developed for sending videoconferencing data over the Internet. These data rates require a high speed Internet connection either through ISDN, T1, cable modem, DSL, satellite, or a LAN.

Videoconferencing equipment can be configured in two ways: PC-based and standalone. PC systems use the standard desktop PC with either a software or a hardware CODEC configuration. The cameras and microphones that are typically included in a PC configuration cater to use on a desk for up to three people, not to a classroom or conference room setting (Dixon, 2000). Software-based systems do not supply the same quality as a hardware-based system and they are not recommended for teaching or high-performance applications (Dixon, 2000). A standalone system is an autonomous system that operates without a PC. The system includes a high-quality microphone and camera. The video must be displayed on a TV monitor or projector. The camera can be controlled with a remote control; some cameras have a tracking mechanism that automatically focuses the camera on whoever is talking. Standalone systems are easy to operate and are ideal for a classroom or conference room setting. The common commercial standalone systems range from $4,000 to $10,000 (Dixon, 2000). It is also possible to establish multipoint Internet videoconferencing involving more than two sites.

Advantages and Disadvantages

The main advantage of videoconferencing is that it allows real-time, two-way interaction between individuals in different places. Participating parties may be in remote areas and may be separated by large distances. All individuals involved in the educational system from the teachers, students, curriculum developers, and specialists, to the policy makers can participate in a videoconference. It can be used for presentations, teaching sessions, discussion, course delivery (in combination with other media), and student support (Perraton et al., 2002,). During course delivery and lectures, instructors can gauge a student's progress and responsiveness immediately. Instructors can answer questions and provide feedback immediately.

The main drawback for videoconferencing is the cost. For traditional systems, there is a high start-up cost for equipment and technical support must be available for the equipment at each site. Even Internet videoconferencing systems have a relatively high cost. PC-based systems require a PC and a compression card while the standalone systems cost in the thousands. Due to the high set-up cost, videoconferencing sites may not be readily available, especially in remote sites and developing countries. Thus, students may have to travel far to those sites that do use the technology. As well, the proper bandwidth and high-speed Internet access required for video conferencing may not be available and thus add to the overall cost.

TEACHING ATTITUDES USING ICTS

According to the limited studies available, five approaches show some potential for attitude development in TVET. These are a democratic approach, indoctrination approach, group discussion, dramatic involvement, and role modelling. The key issue here is, can ICTs be used for developing attitudes in TVET? This question will remain unanswered for lack of empirical evidence. However, one could speculate about the appropriateness of using ICTs to help students develop desired attitudes.

The indoctrination approach is often used for teaching safety attitudes in TVET. It is possible to indoctrinate learners in an ICT learning environment by creating a very rigid environment, one which does not encourage thinking and where the only choice to students is to learn a set of strict rules. The ICT learning environment can lend itself well to group discussions using bulletin boards and Internet telephony. These group discussions must be carefully moderated to prevent a negative and undesirable attitude from developing. Dramatic involvement can be achieved using games and simulations within an ICT environment. Finally, electronic mentoring could be used to provide role models to learners.

Using a strictly technology-based learning environment eliminates face-to-face contact, hampering the development of people skills that are just as valuable in the workforce as technical skills. Audio or video conferences, online discussion groups, e-mail, etc. do allow interaction with other people; however, these tools are not considered as effective in improving people skills as dealing with someone face-to-face. According to management consultant David Anderson of Anderson and Associates, nothing can replace direct people interaction. "When you're interacting with a Web page you can be as uninterested, sarcastic, or rude as you like. The Web page won't call you on it. People skills need to be taught by people" (Delio, 2000).

The teaching of attitudes remains a challenge for even face-to-face instruction. Attempting to teach attitudes using ICTs is a significantly more challenging task, one plagued with uncertainty. The decision to use or not to use ICTs to teach attitudes in TVET must be guided by a risk assessment focused on the consequences of inadequate performance. Is the risk negligible, serious, or life threatening?

TEACHING PRACTICAL SKILLS USING ICTS

The literature and research dealing with developing practical skills in a face-to-face environment is limited. Romiszowski (1981) argued that: "training of physical skills ...seems to have had a history quite divorced from the main stream of educational research and development. This is probably because for a long time educators tended to look down on the physical skills domain as unworthy of their attention".

In commenting on the use of ODL for teaching practical skills Hampton (2002) noted: "the learning of practical skills is most often associated with workshops and laboratories, specialist materials and equipment, smaller class sizes and, frequently, longer blocks of time for practice and rehearsal. For the open and distance learning (ODL) environment, the teaching of practical skills poses more difficulties that the teaching of knowledge and theory". She added that the teaching of practical skills would always require the support of skilled people and a range of tools or equipment. Haddad and Draxler (2002) noted however that the inability to meet the challenge is self-inflicted because of a tendency to use Industrial Age models to solve Information Age problems.

Oliveira and Rumble (1992) noted that for many years educators believed that ODL was only suitable for teaching cognitive skills. Sparkes (1982) argued that it might be more effective to teach a practical skill using a video rather than face-to-face instructions. He does concede, however, that practical skill will continue to need face-to-face supervision at least until two-way television is available. The author identified three organizations using ODL for teaching performance skills: National Extension College, UK; University of Victoria, Canada; and Indonesia Banking Development.

Analysing these categories sheds light on how practical skills develop and provides some insights regarding the implications of using ODL for teaching such skills. It is interesting to note that only during the imitation stage does the learner need explanation and demonstration. During the manipulation, precision, articulation, and naturalization stages the learner can practice independently to perfect his/her skill. Consequently, the imitation stage is the most critical stage of practical skills development and it is this phase that requires most guidance and support in an ODL environment.

Hampton suggested that it is critical to take into consideration the learner's learning style in choosing or developing materials for teaching practical skills by distance. She noted that print-based illustration and step-by-step procedures are most often used for learning a practical skill at a distance and that video, interactive multimedia on CD-ROM and online learning are increasingly being used. She also assesses these three approaches. Advantages and problems with these three approaches were discussed previously in section two.

Other approaches in current use for developing practical skills using ICTs are online learning, simulators and virtual reality. Pilots, truck drivers, and crane operators may spend more training time in a virtual reality environment and in simulators than with the actual equipment to reduce hazards and equipment rental costs. Majumdar & Ray (no date) states that the military has always been using state-of-the-art technology through simulation packages for teaching dangerous tasks like flight training, astronaut training, surgical operations, and handing nuclear energy. The author noted that the use of ICTs allows these skills to develop without using expensive equipment or subjecting trainees to extreme danger. In addition to technological support, appropriate learning environments must be provided to enable learners to practice and perfect their skills. The types of support that have been proved to be effective include: mobile workshops, learning centres or learning commons, workplace training, on-the-job training, work-based learning, and work experience (Hampton, 2002).

Assessing skills acquired by distance learning is as challenging as developing them. The assessment of practical skills involves two elements: process – formative evaluation, and product – summative evaluation. Process assessment involves evaluating the procedures used for executing a practical task. Product assessment, as the name suggests, consists of evaluating the quality, accuracy, and finish of a final product. Hampton (2002) suggests that process evaluation can be accomplished through self-assessment and that product evaluation can be achieved using demonstration of skills, video recording of performance, validators, and workplace trainers and assessors.

USING ICTS FOR WORKPLACE TRAINING

The Conference Board of Canada (2001a) notes: "more than ever, employers depend on knowledgeable and skilful employees to create value-added products and services, efficiently and effectively, so that they can compete successfully". Economic prosperity in the global economy is becoming highly dependent on an ability to develop, attract, and maintain a well-educated workforce. As a result of these changes, the demand for skilled workers often exceeds the supply in many sectors of the economy. Seventy per cent of Fortune 1000 companies perceive the lack of a pool of talented workers as the biggest barriers to sustaining growth. In order to ease this skill gap, employers are significantly investing in workplace education and training.

Trends indicate that workers acquire 70% of the skills they need for the workplace through informal training. Employers are increasingly embracing e-learning solutions to meet this growing requirement for workplace training. It appears that information on employers' involvement in e-learning may not be readily available. Wolley and Booker (no date) expressed difficulties in researching the use of ICTs for corporate training in Australia.

In its recent survey of 29 countries with advanced economies, OECD found the two most important incentives for integrating e-learning in the corporate world were reduced costs and increased flexibility (OECD, 1999). Bates (2001) noted that the most dramatic impact of ICTs has been on campus learning and corporate training. Canadian employers intend to increase their use of e-learning from 17% to 24%. This is consistent with American counterparts who also foresee a growth in the use of ICTs for workplace learning – from 19.8% to 24%. It is estimated that 50% – 70% of American firms use the web for instructional delivery and that corporate spending for online learning in that country will reach $11.4 billion (Zirkle, 2002). The estimated growth of e-learning for 2003 in the corporate sector and higher education, combined, is in the area of $18 billion (Bates, 2001). A significant proportion of employers are using both the Internet and Intranet to meet their learning needs. Research conducted by the Conference Board of Canada (2001a) indicates that the top reasons given by employers for using ICTs in training and development are: improved just-in-time workplace learning (60%), cost-effectiveness (45%), greater control of employees over learning (35%), and best fit with organization's workplace learning needs (33%).

In recent years there has been a proliferation of "corporate universities" in the private sector. These training facilities make extensive use of ICTs such as videoconferencing and the Internet for teaching and learning. The unprecedented demand for workplace training has also encouraged the development of an online learning industry led by contractors interested in servicing the training needs of small- and medium-sized companies (Bates, 2001). Fifty-one per cent of Canadian employers' e-learning solutions are developed in-house, while 49% are acquired off-the-shelf (Conference Board of Canada, 2001). Although corporate e-learning is growing at an exponential rate and is perceived by many employers as an essential tool for keeping their employees workplace-ready, there are serious barriers that hinder e-learning implementation. The top three starting barriers are: cost of learning technologies, time required for programme development, and lack of content in learning technology format. The top three implementing barriers include: cost of learning technologies, lack of dedicated learning time, and time necessary for developing e-learning programmes (Conference Board of Canada, 2001).

Using ICTs for Assessing Teaching and Learning

Three broad strategies are used in assessing performance in TVET. Cognitive skills are assessed using written tests, practical skills are assessed for process and product, and attitude is assessed through ongoing observation. ICTs can provide considerable support to TVET instructors in the assessment process. A computer with the associated software can be used to design, develop, administer, score, and evaluate objective tests and examination. A hand-held computer can be used for the progressive assessment of practical work or to record observation data in the assessment of attitudes. Electronic records of performance allow instructors to monitor students' progress regularly and identify patterns and trends that would lead to improved teaching. ICTs can also be used to assist students in generating an electronic portfolio.

Using ICTs for Prior Learning Assessment and Recognition

The Conference Board of Canada (2001b) notes: "more than ever, employers depend on knowledgeable and skilful employees to create value-added products and services, efficiently and effectively, so that they can compete successfully". Economic prosperity in the global economy is becoming highly dependent on its ability to develop, attract, and maintain a well-educated workforce. As a result of these changes, the demand for skilled workers often exceeds the supply in many sectors of the economy. In order to ease this skill gap, there has been an increased interest in developing and implementing policies and practices for assessing and recognizing prior learning. The process is known as Prior Learning Assessment and Recognition (PLAR). Although adults spend a considerable amount of time on employment-related training, educational institutions fail to recognize knowledge and skills acquired outside of formal education environments. A recent report issued by the Council of Ministers of Education of Canada (1998) suggests that the full range of knowledge and skills of Canadians should be recognized and used efficiently and comprehensively in Canada's education systems and workplaces.

ICTs can be used in TVET to provide new and innovative strategies to facilitate the assessment and recognition of prior learning. Alboim (2002) presented a very comprehensive model of PLAR involving the use of ICTs at the Canadian National Summit on Innovation and Learning. Her model is primarily designed to facilitate the labour market entry of skilled immigrants in Canada. The main role of ICTs in this model is through an Internet portal.

Some key components of this proposed portal include:

- Information;
- Assessment services for academic achievement; technical skills; language and Canadian work practices;
- Labour market counselling and learning plans; and mentorship by Canadian practitioners.

ICT LITERACY SKILLS FOR VOCATIONAL DEVELOPMENT

ICT LITERACY FOR TVET TEACHERS

It appears that ICTs have made little impact in schools in spite of the seemingly great potential to improve teaching and learning (Miller, 1997). Dyril and Kinnauman (1994) noted, "Technology has transformed every segment of American society – except education...schooling today remains much the same as it was before the advent of personal computers. The lack or inadequate training of teachers is considered to be one the major barriers for the integration of ICTs in TVET (Miller, 1997). Cuneo and associates (2000) argued that lack of staff training in ICTs is a major cause of the digital divide among Canadian post-secondary institutions. Research indicates that the computer literacy of TVET teachers can decrease computer anxiety (Birkenholz and Stewart, 1991) and improve attitudes towards computers (Fletcher and Deeds, 1991). On the basis of these findings Fletcher and Deeds (1991) recommended implementing additional pre-service and in-service ICT training for TVET teachers. According to Allen, Walker, & Morehead (1999), "Many educators are limited in the technological skills needed for successful technology integration in career ad technical education curriculum due to the lack of recent training in the professional field".

There is now a general consensus that teachers are the principal agents for integrating technology in education and training. A study conducted by Marie Jasinski (1998) involving 80 TVET educators indicated that technology does not improve or cause changes in learning. It is well-designed instructions and how teachers make use of the new capabilities, potentials, and options offered by ICTs that can improve learning significantly.

In spite of the importance of ICT literacy skills for TVET teachers, it appears that they are not being adequately prepared to meet the challenge. A study conducted by Miller (1997) indicated that only 50% of teacher education institutions surveyed had implemented a required course in ICTs. The most striking finding of this study is that computers were most frequently being used as "glorified typewriters" in vocational teacher education. A forum on trainer training and e-learning hosted by CEDEFOP Electronic Training Village (ETV) in 2001 indicated that "basic technology skills are often acquired by trial and error learning in day-to-day teaching practice" (ANTA, 2001a).

In considering ICT literacy skills for TVET teachers two main skill dimensions must be considered: technical and pedagogical ICT literacy, and occupational ICT literacy. Technical and pedagogical ICT literacy refers to the technical literacy skills for the equipment being used and the specialized skills necessary in its pedagogical application. Occupational ICT literacy refers to the competencies related to the use of technology-based equipment and control systems.

Technical and Pedagogical ICT Literacy for TVET Teachers

The International Training Centre of the ILO (2002) in Turin, Italy has developed a set of Training Technology Competence Standards for TVET personnel involved in distance education and learning technology applications. This comprehensive competency profile includes four major standards and 24 subcategories, as well as specific performance indicators for each. The case study that follows outlines these competencies.

4

Usability Valuation Techniques for Interactive Television

INTRODUCTION

Governments and businesses are interested in using the penetration of the TV set into Western homes to sell products and services, disseminate information, enable learning and so on via interactive TV services and facilities. "Interactive" in the television context is understood to include video on demand, t-commerce, interactive programme guides and enhanced broadcast television. However, if they are to be taken up, interactive facilities need to be usable by viewers. This brings new challenges for television programme producers, who do not have a strong tradition of fine grained analysis of viewer interaction with television, preferring instead to rely on survey methods such as diaries, questionnaires, focus groups or automated monitoring to discover viewers' attitudes (Gauntlett & Hill, 1999).

Specialists in user interface evaluation, on the other hand, are well acquainted with the issues involved in testing a range of applications designed for desktop PC's, but are now faced with the prospect of modifying their techniques and methods to cope with the demands of new platforms, including interactive television. Several evaluation techniques may be applicable to iTV, including analytical approaches such as heuristic evaluation (Nielsen, 1993). However it is not certain that general desktop-oriented heuristics such as "speak the user's language", "be consistent" and so on still apply in the same way and to the same extent. Building on the growing evidence from studies reported in the literature it is beginning to be possible to derive TV-specific heuristics (Daly-Jones, 2002).

However, nothing is very strongly established yet. Other analytical techniques, such as Cognitive Walkthrough (Newman & Lamming, 1995) may also be useful and are probably a very effective technique to use during the design and development process, particularly of task-oriented elements. However, here we concentrate on empirical evaluation, based on observation and interview sessions with viewers.

WHAT IS INTERACTIVE TELEVISION

Interactive television is a person-to-person technique that allows two-way communication. Unlike conventional one-way television (TV) or radio broadcasts, most interactive TV enables viewers to respond by voice telephone or computer connected to an appropriate hosting service (Internet Service Provider, special on-line bulletin board, chat room, etc.). A further refinement of the technology uses sophisticated equipment, TV cameras, and special connections at both ends so that participants can see and hear one another. This kind of interactive TV is usually limited to small groups for long-distance conferences.

Interactive television is characterized as follows:

- A television broadcast includes telephone numbers or computer addresses to use in responding;
- Participants use telephones or computers to comment or ask questions; and
- Staff is available to record comments or respond to questions.

Electronic town meetings are a good example, because large numbers of people participate directly from their homes or other designated locations. A meeting, presentation, or panel discussion is held in a central location with an audience, while a TV crew records and broadcasts the proceedings over local cable. Home viewers phone in questions for discussion leaders to answer—a format similar to a talk-radio call-in program. The Southern California Association of Governments uses interactive television to reduce the distance the public has to travel to participate in meetings. Conferencing equipment is placed at central locations in each of the six counties it serves.

Why is it Useful

"A picture is worth a thousand words." Interactive TV provides direct or immediate knowledge of what transpired at a public meeting, which is useful for some people compared with a written summary document. Interactive TV helps people grasp a planning concept, understand complex programs, and absorb large amounts of information quickly. Television is an integral part of many people's lives. It attracts broader participation in a public involvement program. Electronic town meetings, for instance, may actively engage "couch potatoes" who would otherwise not participate in civic affairs.

Electronic town meetings increase awareness about a project or program. They are very useful for developing consensus across a broad range of participants. They provide a large segment of the population with direct, timely access to key decision-makers.

Interactive TV offers the immediacy of a "live" broadcast. During a broadcast, participants at home respond or participate via telephone. They then see and hear that their concerns are being addressed, and possibly respond further. Town meeting audiences can convene in several locations. With interactive TV, large groups may take part at a central location while numerous individuals or small groups participate from homes or satellite meeting halls. The Central Puget Sound Regional Transit Authority (RTA) in Seattle held a Satellite Summit prior to adopting a proposed ballot measure on a major new transit system. From a central studio, the RTA board addressed audiences at five remote locations as well as home viewers via cable TV. Audience members at remote sites were able to pose questions and offer suggestions directly to the board. A videotape of the event was sent to local libraries for reference by those people unable to participate at broadcast time.

Informal surveys can be a central element in interactive TV. Viewers use telephones to register approval or disapproval on a specific project or issue under discussion. Results are tabulated and shown on the program, perhaps generating additional responses.

Does it have Special Uses

Interactive TV is especially useful for public presentations. Information can be disseminated at regular intervals to the audience at their homes. Interactive television can target a specific audience. Broadcasts can reach specialized audiences through non-English language or other special media channels or shows. Berks Community Television in Reading, Pennsylvania, is a two-way cable television program designed to reach seniors, a special use of the technology for a targeted audience.

A broadcast helps an agency reach a wider audience than it might otherwise find. It increases both awareness and inclusiveness. Traditional broadcast technology is used for limited educational and outreach purposes, but the incorporation of two-way communication through interactive technology expands the participatory aspect of the medium. To increase public participation in regional transportation planning, Orange County replaced a public hearing with a "Community Dialogue," sponsored by the Huntington Beach City Council Chamber.

A talk-show format was used to interview transportation and land use planners about the impact of growth on the future of Southern California. Questions from an in-studio audience and call-in viewers were answered live on television. The show was run from 7:00-9:00 PM on a weeknight and simultaneously cablecast in nine different cities close to Huntington Beach. In addition, the cable company aired the program two more times during the day. The cable company estimated that more than 7,500 people watched the show.

A broadcast allows instant feedback that can be shared with the entire community. Because people participate from their own homes, they do not need to arrange for childcare or worry about transportation or proper appearance.

Who Participates? And how?

An interactive TV program involves many people—particularly if the program is well publicized. Broadcasts on a major local or regional channel stand the best chance of reaching many viewers for a program on transportation planning or project development.

Cable TV subscribers are major users. Many public agencies have access to cable TV channels over which electronic town meetings may be carried. Any local person with access to such cable services and a telephone line may participate. Participation is limited in areas without cable access or if people do not subscribe to the service. Viewers call the TV station during a broadcast to register an opinion or comment. Viewers may call repeatedly if they do not receive a response or if they feel that the response is inadequate.

Viewer responses are recorded on the voice track of a videotape of a program. As a program is being taped, a viewer calls in and converses with an agency person at the TV studio. The video equipment picks up both ends of the conversation. Responses may also be recorded, either automatically by an answering device or in person by agency staff answering telephone calls and recording responses in writing or on computer.

Viewer input may provide immediate feedback. Calls from viewers inform the studio and home audiences of preference responses, survey results, or participant concerns.

How do Agencies Use the Output

Viewer comments help an agency gauge the level of community interest and concern about transportation issues. Heated, lively debate and strongly worded comments signal a controversial topic and indicate that the agency needs to reach out in other ways to the full array of concerned parties.

Viewer feedback also helps an agency identify community perceptions about critical issues, impacts that are most sensitive, alternatives that are preferred or proposed, and ways to improve plans and make responsive decisions. In metropolitan planning, an agency can use viewers' comments to identify differences of opinion and needs in subareas of a region or among types of interest groups. To get a clear sense of how public opinion is changing over the course of a project, a record of comments serves as a benchmark that can be compared with past or future responses. Such reference points assist an agency in evaluating long-term program goals or objectives and reassessing meeting techniques.

Viewer input may be used to expand mailing lists. To increase the availability of transportation information, for instance, names and addresses of respondents are registered along with their questions and concerns.

Who Leads

Skilled professionals are required. Interactive TV demands the special technical skills of studio crews and facilitators who coordinate feedback through a telephone company. Although conventional presentation needs and requirements of a public meeting come into play, a skilled moderator for the program is also required.

Agencies sponsor regular programs. Vermont Interactive Television shows live, two-way programs on subjects such as early childhood and family support meetings, technical mathematics, and Vermont history and government. It combines education with participatory technologies to bring together thousands of participants throughout a largely rural State.

What are the Costs

Interactive TV such as an electronic town meeting is expensive. The sophisticated equipment and skilled operator requirements raise costs. Contracting services are necessary for high-speed digital telephone lines to accommodate incoming calls and instantly tabulate data. An agency with access to a public TV channel may be able to reduce costs. For example, community access cable channels and/or schools with broadcast media facilities, both of which have public service missions, may be able to be used.

How is Interactive Television Organized

Cooperation from a local TV station is essential. A local station such as a local access cable channel or a local government or college station reaches most, if not all, homes in a broad area. A broadcast or town meeting event must be publicized on the station as well as in other print and electronic media so that people know when to tune in. A local TV station may also recommend a prominent media personality to moderate the program, keep discussion moving, and enliven the program. University stations might include student production of public meetings on interactive TV as part of the curriculum.

Technical assistance on broadcasting is essential. Local stations are likely to be conversant with interactive techniques or can help find the right contacts.

Comparison with a public meeting or hearing is useful as a starting point. Many public agencies are versed in the logistical requirements and strategic use of such meetings. A presentation of essential facts about the project or program provides a springboard from which discussion takes place. However, incorporation of interaction affects the program format and shapes the agenda to reach out to a larger audience.

How is it Used with Other Techniques

Interactive TV supplements a broader outreach program. It cannot be an agency's sole means of communication with the public. Due to costs and time constraints, interactive TV may accent or bring to culmination a larger effort to inform the public. However, it also can be useful as a survey of public reactions to an issue. It can function as a large-scale public meeting. Savannah,

Georgia, broadcasts several meetings on the same topic from different locations with trained facilitators and programmers. Home viewers call in to respond, adding a sense of dynamic energy to the process.

What are the Drawbacks

Perceptions of the meaningfulness of participation via TV vary. Participants may doubt that a TV program has lasting impact. As is true with any new technology, people sometimes resent its use, because they perceive it as a replacement for personal communication.

Imbalance is magnified by live TV. With any project or program, the danger arises that only one or a few interests will participate and that the dialogue will not accurately reflect the full array or relative strength of community opinions. Callers may want to grandstand a particular issue. Agencies can limit speaking times but cannot deny a determined individual the opportunity to speak. This emphasizes the importance of agency outreach to the full array of community interests.

Broadcast adds pressure for quick decisions. Television events often put decision-makers in direct contact with members of the community, and the community may want immediate decisions. One transit agency held an open house for a major investment study that was heavily attended by a community that opposed one alternative. The community group arranged for the meeting with agency and legislative leaders to be broadcast in a hearing format over a local TV channel. During the meeting, a speaker (with vocal support from the crowd) demanded a satisfactory decision from the agency and the legislators in two weeks, rather than the several months scheduled in the study plan.

Input from interactive TV, like that from informal surveys, is not statistically representative. Only interested people participate, and cable viewership in many areas is minimal. Broadcast responses supplement but do not substitute for more formal survey data as an accurate way to gauge public reaction.

Is Interactive Television Flexible

Interactive television is inherently flexible for verbal presentations. An agency spokesperson agency can update the community on a project or a planning process. Adding graphics to a presentation, however, requires additional time and effort. Interactive TV is as flexible as other public meeting formats. However, the sponsoring agency is limited to times available on a station's program schedule.

When is it Used most Effectively

Electronic town meetings are most effective at an important juncture when focused, relevant public input is needed. The Central Puget Sound RTA, in partnership with a local commercial station, presented a two-hour, prime time program on its proposed rapid transit system. Moderated by a local media personality, the program showed features of transit systems from other cities. A 100-person audience was able to show approval or disapproval for various options via hand-held opinion meters that scored opinions on a scale of one to ten. A panel of experts and critics kept the discussion balanced.

Interactive television can continually update transportation information. A cable TV program can offer a telephone number to call for further information. Its use is especially valuable for conveying images or visual representations of ideas, including renderings and animations of existing and potential conditions.

Interactive television can build momentum for or against an improvement. This is particularly important when funding sources are in question. However, project opponents can also "hijack" that momentum. Political leaders who make budget decisions pay attention to high-profile events that reach a broad segment of their constituencies.

FORMS OF INTERACTION

The term "interactive television" is used to refer to a variety of rather different kinds of interactivity (both as to usage and as to technology), and this can lead to considerable misunderstanding. At least three very different levels are important (see also the instructional video literature which has described levels of interactivity in computer-based instruction which will look very much like tomorrow's interactive television):

Interactivity with a TV Set

The simplest, Interactivity with a TV set is already very common, starting with the use of the remote control to enable channel surfing behaviors, and evolving to includevideo-on-demand, VCR-like pause, rewind, and fast forward, and DVRs, commercial skipping and the like. It does not change any content or its inherent linearity, only how users control the viewing of that content. DVRs allow users to time shift content in a way that is impractical with VHS. Though this form of interactive TV is not insignificant, critics claim that saying that using a remote control to turn TV sets on and off makes television interactive is like saying turning the pages of a book makes the book interactive.

In the not too distant future, the questioning of what is real interaction with the TV will be difficult. Panasonic already has face recognition technology implemented its prototype Panasonic Life Wall. The Life Wall is literally a wall in your house that doubles as a screen. Panasonic uses their face recognition technology to follow the viewer around the room, adjusting its screen size according to the viewers distance from the wall. Its goal is to give the viewer the best seat in the house, regardless of location. The concept was released at Panasonic Consumer Electronics Show in 2008. Its anticipated release date is unknown, but it can be assumed technology like this will not remain hidden for long.

Interactivity with TV Program Content

In its deepest sense, Interactivity with normal TV program content is the one that is "interactive TV", but it is also the most challenging to produce. This is the idea that the program, itself, might change based on viewer input. Advanced forms, which still have uncertain prospect for becoming mainstream, include dramas where viewers get to choose or influence plot details and endings.

- As an example, in Accidental Lovers viewers can send mobile text messages to the broadcast and the plot transforms on the basis of the keywords picked from the messages.
- Global Television Network offers a multi-monitor interactive game for Big Brother 8 (US) "'In The House'" which allows viewers to predict who will win each competition, who's going home, as well as answering trivia questions and instant recall challenges throughout the live show. Viewers login to the Global website to play, with no downloads required.
- Another kind of example of interactive content is the Hugo game on Television where viewers called the production studio, and were allowed to control the game character in real time using telephone buttons by studio personnel, similar to The Price is Right.
- Another example is the Clickvision Interactive Perception Panel used on news programmes in Britain, a kind of instant clap-o-meter run over the telephone.

Simpler forms, which are enjoying some success, include programs that directly incorporate polls, questions, comments, and other forms of (virtual) audience response back into the show. One example would be Australian media producer Yahoo!7's Fango mobile app, which allows viewers to access program-related polls, discussion groups and (in some cases) input into live programming. During the 2012 Australian Open viewers used the app to suggest questions for commentator Jim Courier to ask players in post-match interviews.

There is much debate as to how effective and popular this kind of truly interactive TV can be. It seems likely that some forms of it will be popular, but that viewing of pre-defined content, with a scripted narrative arc, will remain a major part of the TV experience indefinitely. The United States lags far behind the rest of the developed world in its deployment of interactive television. This is a direct response to the fact that commercial television in the U.S. is not controlled by the government, whereas the vast majority of other countries' television systems are controlled by the government. These "centrally planned" television systems are made interactive by fiat, whereas in the U.S., only some members of the Public Broadcasting System has this capability. Commercial broadcasters and other content providers serving the US market are constrained from adopting advanced interactive technologies because they must serve the desires of their customers, earn a level of return on investment for their investors, and are dependent on the penetration of interactive technology into viewers' homes.

In association with many factors such as:

- Requirements for backward compatibility of TV content formats, form factors and Customer Premise Equipment (CPE)
- The 'cable monopoly' laws that are in force in many communities served by cable TV operators
- Consumer acceptance of the pricing structure for new TV-delivered services. Over the air (broadcast) TV is Free in the US, free of taxes or usage fees.
- Proprietary coding of set top boxes by cable operators and box manufacturers
- The ability to implement 'return path' interaction in rural areas that have low, or no technology infrastructure
- The competition from Internet-based content and service providers for the consumers' attention and budget
- And many other technical and business road blocks.

Interactivity with TV-related Content

The least understood, Interactivity with TV related content may have most promise to alter how we watch TV over the next decade. Examples include getting more information about what is on the TV, weather, sports, movies, news, or the like.

Similar (and most likely to pay the bills), getting more information about what is being advertised, and the ability to buy it—(after futuristic innovators make it) is called "tcommerce" (short for "television commerce"). Partial steps in this direction are already becoming a mass phenomenon, as Web sites and mobile phone services coordinate with TV programs (note: this type of interactive TV is currently being called "participation TV" and GSN and TBS are proponents of it). This kind of multitasking is already happening on large scale—but there is currently little or no automated support for relating that secondary interaction to what is on the TV compared to other forms of interactive TV.

Others argue that this is more a "web-enhanced" television viewing than interactive TV. In the coming months and years, there will be no need to have both a computer and a TV set for interactive television as the interactive content will be built into the system via the next generation of set-top boxes.

However, set-top-boxes have yet to get a strong foothold in American households as price (pay per service pricing model) and lack of interactive content have failed to justify their cost.

One individual who is working to radically disrupt this field is Michael McCarty, who is the Founder and CEO of a new wave of interactive TV products that will be hitting the market in early 2013. As he suggested in his presentation to the "Community for Interactive Media", "Static media is on its way out, and if Networks would like to stay in the game, they must adapt to consumers needs."

Many think of interactive TV primarily in terms of "one-screen" forms that involve interaction on the TV screen, using the remote control, but there is another significant form of interactive TV that makes use of Two-Screen Solutions, such as NanoGaming. In this case, the second screen is typically a PC (personal computer) connected to a Web site application. Web applications may be synchronized with the TV broadcast, or be regular websites that provide supplementary content to the live broadcast, either in the form of information, or as interactive game or program. Some two-screen applications allow for interaction from a mobile device (phone or PDA), that run "in synch" with the show.

Such services are sometimes called "Enhanced TV," but this term is in decline, being seen as anachronistic and misused occasionally. (Note: "Enhanced TV" originated in the mid-late 1990s as a term that some hoped would replace the umbrella term of "interactive TV" due to the negative associations "interactive TV" carried because of the way companies and the news media over-hyped its potential in the early 90's.)

Notable Two-Screen Solutions have been offered for specific popular programs by many US broadcast TV networks. Today, two-screen interactive TV is called either 2-screen (for short) or "Synchronized TV" and is widely deployed around the US by national broadcasters with the help of technology offerings from certain companies.

One-screen interactive TV generally requires special support in the set-top box, but Two-Screen Solutions, synchronized interactive TV applications generally do not, relying instead on Internet or mobile phone servers to coordinate with the TV and are most often free to the user. Developments from 2006 onwards indicate that the mobile phone can be used for seamless authentication through Bluetooth, explicit authentication through Near Field Communication. Through such an authentication it will be possible to provide personalised services to the mobile phone.

Interactive TV Services

Notable interactive TV services are:

- *ActiveVideo Networks (formerly known as ICTV)*: Pioneers in interactive TV and creators of CloudTV™: A cloud-based interactive TV platform built on current web and television standards. The network-centric approach provides for the bulk of application and video processing to be done in the cloud, and delivers a standard MPEG stream to virtually any digital set-top box, web-connected TV or media device.

- *T-commerce*: Is a commerce transaction through the set top box return path connection.
- BBC Red Button
- *ATVEF*: 'Advanced Television Enhancement Forum' is a group of companies that are set up to create HTML based TV products and services. ATVEF's work has resulted in an Enhanced Content Specification which makes it possible for developers to create their content once and have it display properly on any compliant receiver.
- *MSN TV*: MSN TV supplies computerless Internet access. It requires a set-top box that sells for $100 to $200, with a monthly access fee.
- *Philips Net TV*: Solution to view Internet content designed for TV; directly integrated inside the TV set. No extra subscription costs or hardware costs involved.
- *Interactive TV purchasing system*: An Interactive TV purchasing system was introduced in 1994 in France. The system was using a regular TV set connected together with a regular antenna and the Internet for feedback. A demo has shown the possibility of immediate purchasing, interactively with displayed contents.
- BrightLine delivers interactive television advertising campaigns to over 90 million households. The company's platform agnostic solutions are launched across all available television platforms, including cable and satellite television providers, telecom companies, gaming consoles, connected televisions, smartphones and tablets such as the Apple iPad. BrightLine's active distribution channels include,but are not limited to: DirecTV, Dish, Xbox Live, PlayStation 3, Cablevision, AT&T U-verse, Verizon FiOS, LG and Samsung SmartTVs, as well as IOS and Android mobile phones and tablets. Clients including American Express, Cisco Systems, Inc., Jim Beam Global, Johnson & Johnson, PepsiCo, L'Oreal, Unilever and Verizon Wireless.

Closed-circuit Interactive Television

Television sets can also be used as computer displays or for video games

USER INTERACTION

Interactive TV has been described in human-computer interaction research as "lean back" interaction, as users are typically relaxing in the living room environment with a remote control in one hand. This is a very simplistic definition of interactive television that is less and less descriptive of interactive television services that are in various stages of market introduction. This is in contrast to the descriptor of personal computer-oriented "lean forward" experience of a keyboard, mouse and monitor. This description is becoming more distracting than useful as video game users, for example, don't lean forward while they are playing video games on their television sets, a precursor to interactive TV.

A more useful mechanism for categorizing the differences between PC and TV based user interaction is by measuring the distance the user is from the Device. Typically a TV viewer is "leaning back" in their sofa, using only a Remote Control as a means of interaction. While a PC user is 2 ft or 3 ft (60 or 100 cm) from his high resolution screen using a mouse and keyboard. The demands of distance, and user input devices, requires the application's look and feel to be designed differently. Thus Interactive TV applications are often designed for the "10-foot user interface" while PC applications and web pages are designed for the "3ft user experience". This style of interface design rather than the "lean back or lean forward" model is what truly distinguishes Interactive TV from the web or PC. However even this mechanism is changing because there is at least one web-based service which allows you to watch internet television on a PC with a wireless remote control.

In the case of Two-Screen Solutions Interactive TV, the distinctions of "lean-back" and "lean-forward" interaction become more and more indistinguishable. There has been a growing proclivity to media multitasking, in which multiple media devices are used simultaneously (especially among younger viewers). This has increased interest in two-screen services, and is creating a new level of multitasking in interactive TV. In addition, video is now ubiquitous on the web, so research can now be done to see if there is anything left to the notion of "lean back" "versus" "lean forward" uses of interactive television.

For one-screen services, interactivity is supplied by the manipulation of the API of the particular software installed on a set-top box, referred to as 'middleware' due to its intermediary position in the operating environment. Software programs are broadcast to the set-top box in a 'carousel'. On UK DTT (Freeview uses ETSI based MHEG-5), and Sky's DTH platform uses ETSI based WTVML in DVB-MHP systems and for OCAP, this is a DSM-CC Object Carousel.

The set-top box can then load and execute the application. In the UK this is typically done by a viewer pressing a "trigger" button on their remote control.

Interactive TV Sites have the requirement to deliver interactivity directly from internet servers, and therefore need the set-top box's middleware to support some sort of TV Browser, content translation system or content rendering system. Middleware examples like Liberate are based on a version of HTML/JavaScript and have rendering capabilities built in, while others such as OpenTV and DVB-MHP can load microbrowsers and applications to deliver content from TV Sites. In October 2008, the ITU's J.201 paper on interoperability of TV Sites recommended authoring using ETSI WTVML to achieve interoperability by allowing dynamic TV Site to be automatically translated into various TV dialects of HTML/JavaScript, while maintaining compatibility with middlewares such as MHP and OpenTV via native WTVML microbrowsers.

Typically the distribution system for Standard Definition digital TV is based on the MPEG-2 specification, while High Definition distribution is likely to be based on the MPEG-4meaning that the delivery of HD often requires a new device or set-top box, which typically are then also able to decode Internet Video via broadband return paths. Emergent approaches such as the Fango app have utilised mobile apps on smartphones and tablet devices to present viewers with a hybrid experience across multiple devices, rather than requiring dedicated hardware support.

DIFFERENT OF INTERACTIVE TELEVISION

There are a number of areas that distinguish the use of personal computers from the use of iTV. These differences suggest that evaluating iTV might need a different approach from desktop applications. They also suggest that results that are reliable for desktop applications may need handling with more caution in an iTV context.

Physical characteristics of interaction: Viewers watch television at some distance from the screen, typically in an environment that is oriented toward relaxation and comfort. Resolution is much lower than computer display screens and colour behaves differently. Detailed information is presented via audio. All interactions are carried out via a handset (combined in some cases with a keyboard). A lab set-up is needed in which views of the handset, screen and viewer's focus of attention can be captured together with any soundtrack.

Multiple information channels are mediated via the same device: There is conflict between watching the broadcast stream and manipulation of any interactive components – viewers must divide their cognitive resources between watching and interacting, and this may be reflected in the design through allocation of screen 'real-estate'. Although an interaction element may be usable in a lab setting when the broadcast stream is not present, it may not work so well which when the viewer's attention is engaged on the broadcast stream.

Embedded nature of services: Services tend not to be stand-alone but embedded within a programme or at least a channel. The perceived usability of a service and aesthetic preference for the embedding programme correlate. The services may in principle be very usable, but if a viewer would never think of watching the programme they enhance, this finding becomes irrelevant. Interaction complexity that may be acceptable to a committed fan may deter the casual viewer.

Broadcast-related aspects: While some services such as EPG's and video on demand can be evaluated fairly easily because they are not particularly time-dependent, enhanced broadcast TV is very difficult to test before broadcast. Motivation for using the interactivity may depend on simultaneous interaction with other viewers, e.g. a tendency to interact in voting application may be heightened if viewers are told a large number of other viewers are sending in votes. The communal experience of watching with potentially millions of others will not be available in non-real-time viewing. Even for summative evaluation of a broadcast product there are problems. The programme may go out during evenings or weekends when usability labs are not normally accessible. Moreover, only one trial per broadcast can be carried out in a single lab.

Time-related aspects: The motivation to carry out a particular interaction may be determined by an arbitrary element of the broadcast. For instance, while watching a football match, a change to a preferred type of view may be required during particular aspects of play such as penalty shootouts. There may be very tight time constraints on initiating and receiving this change. The continuity of monitoring the event being presented must not be lost and at any time the change may be immediately cancelled or modified. Thus there are characteristics in common with real-time control systems. Exercising this interactivity should be at the spontaneous whim of the viewer. Instructing the test subject to use this option may mean that significant aspects of it are missed.

The optional status of television viewing: Television tends to mean leisure and entertainment rather than work or other serious pursuit. Almost by definition it is never going to be mandatory for anyone to use interactive services in the way that using a work-oriented application such as database or even a word processor might be (though exceptions might exist, e.g. for educational TV). Thus the task-oriented approach most often adopted by usability evaluators may be inappropriate. Even when a task, such as booking a holiday or checking a viewing time, is the focus, viewers always have a choice of accomplishing it via an alternative means to the television, e.g. by using the telephone or consulting a newspaper. This means that assumptions about subjects' willingness to persevere with a task in non-controlled circumstances need to be very conservative.

Social characteristics of the interaction: The domestic setting in which TV is consumed is complex and is difficult to emulate in all its facets. People have a tendency to do other things, such as ironing, knitting or reading while viewing. They often view in company. They may be subject to interruptions of varying frequency and significance. Use of TV interactivity may compete with other domestic resources, e.g., for digital satellite in the UK, the telephone line. Users may spend a long time using a back channel in the lab that they'd be unlikely to do at home, due to cost or pressure from other family members.

The economics of viewing: Many TV interactions have a billable cost associated with them (often the reason for their existence). Who is paying the bill is a big issue that will impact on decision as to if and when to interact and may influence the options exercised. Simulating this is the lab is extremely difficult.

WHAT ARTEFACTS CAN WE EVALUATE

As in any design project, we need some sort of artefact, design or prototype before evaluation is possible. With broadcast iTV programmes, however, this is not possible. There is only one shot, which means that any evaluation must be done at broadcast time and necessarily be summative in nature. This is much less flexible even than conventional TV where a video can be used to time-shift. Interactive elements cannot be recorded and played back in this way.

If we are not to evaluate iTV at broadcast time, the interactivity must be simulated using some kind of prototype. There is some scope for using paper prototyping, particularly for the task-related elements of a service, e.g. t-commerce transactions, an EPG search or an account-related task such as changing a PIN number. The televisual aspects of the product would be lost, but this is often the case anyway in paper prototyping, where high fidelity is not the goal. 'In principle' usability for the task can be established. Other options are prototypes in HTML, an authoring package such as Director or the target middleware for the application. All these options are relatively labour intensive, particularly some varieties of middleware. In addition there is a danger of the prototype working much more rapidly than the delivered service (due to absence of transmission delays) so it may be necessary to integrate pauses to simulate realistic performance.

Selection of Participants

Once a prototype is available, participants need to be recruited. Participant selection may be quite complex. For many desktop Windows™ based applications, it is sufficient to choose testers with broadly similar occupational backgrounds. For iTV, the experience level will vary widely and often be confined to one specific system, e.g. in the UK, Freeview, Sky, NTL or Telewest. Experience of and attitudes to other technologies such as mobile phones, Internet and Teletext may influence performance. Subjects should obviously be from the target demographic group for the programme or service. In addition, ideally subjects should be selected who would normally view together. It is not trivial to compose a group of family or friends, particularly if the observation session must take place at the time the relevant programme is aired.

Facilities for Observation

Clearly, a conventional office-type usability lab will not do for evaluating iTV. At the University of Brighton we have set up the Domestic Technology Usability Lab for evaluation of usability and accessibility aspects of domestic multimedia technology, particularly interactive television. The lab consists of a room, four metres by three metres, fitted out as a domestic lounge. It contains two and three-seat sofas, a coffee table, standard lamp and a wide screen domestic television. Set in a side wall is a two-way mirror enabling the occupants of the room to be observed without distraction. Close-circuit television cameras on motorised mounts within the room enable recording of the occupants. The television display may also be recorded directly. A microphone in the room picks up conversation.

The video and audio feeds from the room are combined via a video multiplexer and digitiser to be stored on the hard disc of a PC. The format of the recorded video can be varied to suit the requirements of the test being carried out, individual stream, picture-in-picture and quad being supported. A simultaneous recording onto VHS tape can be made as a backup. Behavioural analysis software is available on the PC. This enables the marking up of the captured video stream so that frequency and timings of significant events may be obtained. To evaluate interactive television use, two cameras and the display are usually recorded in quad format. A camera above the TV set records a panoramic view of the room, showing the positions and interactions of the room's occupants. A second camera fitted with a motorised zoom lens, placed to one side of the subjects' position, records the manipulation of the TV remote control. The motorised pan, tilt and zoom provide a constant close up view of the remote enabling individual button pushes and other manipulations to be recorded.

Practical Issues

The most effective approach we have found to be co-discovery, where pairs or trios of users, ideally in family or friendship groups, are observed together. This makes "thinking aloud" a natural activity, allowing decision making and other processes to be captured. One needs to be aware, however, that the dynamics of the group may effect the interaction. In particular, control of the handset in couples and families may be an issue.

As far as the physical set-up is concerned, we try to encourage viewers to sit within camera shot, close enough together to pass over handset without stretching. This means that handset passing is not prevented. To create a relaxed atmosphere, refreshments need to be available, at no more than arm's length. These are tailored to fit the group: beer for football fans, fizzy drinks for children, tea for older participants and so on. Participants typically prefer to keep their belongings with them, so there is a need to ensure that handbags and coats do not obscure the camera's view of handset use.

Motivation is a major issue. Instructions and tasks need to be "very simple and readily achievable". It often becomes clear that participants are only persisting with a task because of the slightly formal situation in which they find themselves. At home, they would have simply reached for the listings magazine or given up the struggle. We find that with some viewer groups, particularly older and less confident viewers, having a facilitator in the room helps to keep the task going. We find it useful to have a communication channel from the observers to the facilitator in the room via headphones. However, the price of putting a facilitator in the room is creating an even more formal situation. Whoever the participants, it is useful to agree in advance a signal or form of words to mark the point at which they would naturally give up the task.

During the observation session, we find it useful to capture (and subsequently tag) the following behaviours:

- Use of handset, including passing from one participant to another
- Switches of attention between screen and handset
- Participants' noticing, commenting on and activating screen controls
- Interaction between participants, e.g. instructions, suggestions and evaluative comments
- Comments on the interaction relating to learning, recognition, remembering, forgetting and awareness of functionality
- Comments on previous models of interaction such as Teletext or Internet browsers. These are particularly interesting for eliciting mental models and terminology

Wherever reasonable, we interview participants directly after observation sessions. Things happen very quickly in sessions and asking questions afterwards clarifies the reasons for particular actions. However, it is important to be extremely tactful in these situations. Participants may be tired or feel that they have "failed" in some way if they do not complete tasks. It is important not to confront a participant with failure, especially if they have previously claimed familiarity with the technology. Another aspect of post-test interviews is that participants, perhaps to please the facilitator, will tend to give more positive evaluations of the session than are warranted from the success of the interactions, often claiming ease of use where things were clearly difficult.

EARLY TELEVISION: TECHNICAL DEVELOPMENTS

Television was not invented by a single inventor; instead, many people working both together and alone contributed to the evolution of TV. One inventor built upon the work of previous inventors, and parallel systems emerged around the same time. This makes the question of who invented television a complicated one, as many inventors around the world were experimenting with television at the same time. The idea of "seeing at a distance" had been around for along time. As early as the mid-nineteenth century, it was imagined that it would be soon possible. In 1879, George du Maurier drew a cartoon for British magazine Punch that showed a mother and a father watching on the wall of their English home a tennis match in Ceylon in which their daughter was playing. They were also able to speak to her over a long-distance telephone. In 1882, the French artist Albert Robida produced a series of drawings in which moving images were transmitted onto the walls of people's living rooms. One of these drawings showed a teacher giving a math lesson; another showed a ballet being performed; another, a desert war being fought.

The word "television" was first used—and presented to the public—in August 1900, in a paper by Russian scientist Constantin Perskyi at Paris World Exhibition.

MECHANICAL TELEVISION

Although television in the form of a mechanical scanning system developed at full speed in the 1920s, it had a long story. Modern development started with the 1873 discovery by Joseph May and Willoughby Smith that the element selenium was capable of conducting small amounts of electricity in direct response to the amount of light falling on it. Later inventors, such as George Carey in 1877, found that banks of selenium cells placed in a mosaic analogous to the human eye and wired individually, could send the elements of a picture as electrical signals from each cell simultaneously to a bank of lamps that lit in response to electricity. In 1880, Maurice Leblanc and others developed the principle of scanning, or viewing pictures elements successively rather than all at once as in a mosaic device, and transmitting them sequentially over a single circuit. A major refinement was a mechanical device, George Nipkow's scanning discs, capable of scanning and transmitting even a moving picture.

In 1884, the German scientist patented his disc, which had one spiral of holes or apertures spread equally around the outer part of a flat disc. The path each aperture swept out (through the angle between apertures) corresponded to a line in the image. The radial distance of each successive aperture changed in equal steps so that, in one revolution, all the apertures swept out the area of one TV frame. By masking off that area and placing a photocell behind it, we have a television camera. By placing a variable light source (usually a neon) behind a similar disc, we have a television display. With synchronization of camera and display, we have the vision channel for television. The mechanical nature of systems based on Nipkow discs, and the sensitivity and bandwidth of photocell amplifiers, initially constrained the television image to mere tens of lines rather than the hundreds of lines of the electronic systems emerging in the thirties.

Although he had devised the theory of scanning, Nipkow failed to put it into practice. Of the many inventors and experimenters who worked on mechanical television in the 1920's, three made outstanding progress: Herbert Ives of Bell Telephone Laboratories; John Logie Baird, a self taught Scottish inventor; and Charles Francis Jenkins, an American inventor.

Ives was assigned, with the help of substantial funds and staff, to keep AT&T "abreast of the general advances in the art of television." His work culminated with wire transmissions in 1927 of still and moving image pictures over hundred of miles. Bell Labs never promoted its system commercially.

One of the better-known experimenters with mechanical television was Charles Francis Jenkins, a prolific American inventor. In May 1920, at the Toronto meeting of the Society of Motion Picture Engineers, Jenkins introduced his prismatic rings as a device to replace the shutter on a film projector. This invention laid the foundation for his first "radiovision" broadcast. He claimed to have transmitted the earliest moving silhouette images on June 14, 1923, but his first public demonstration of this did not take place until June 1925. Jenkins Laboratories constructed a radiovision transmitter, W3XK, in Washington, D. C. The short-wave station began transmitting radiomovies across the eastern U. S. on a regular basis by July 2, 1928.

Baird is generally credited with establishing television in Great Britain. Baird used a ventriloquist dummy called Stukey for his experiments on Frith Street, since no human was likely to tolerate for hours the extremely bright lighting that Baird needed in his efforts to obtain a decent picture. On October 1, 1925, Baird managed to get Stukey's head to appear on screen. Inspired by this achievement, he arranged a demonstration for members of the Royal Institution on January 26, 1926. Turning next to live action scenes as a source, in February 1928 he televised a woman's image from London to New York. Later that year he transmitted to the liner Berangaria a thousand miles at sea. And in 1932, he transmitted the English Derby to a large screen in a London movie theater, where 4000 people watched the race. This was similar to what Alexanderson was doing in Schenectady in 1930.

In 1935, when the Television Committee of the British government had to decide which television standards they would use, comparative tests were held of the Baird mechanical system and an electronic system controlled by Electrical and Musical Industries Ltd. (EMI), a company based partly on Zworykin's work in the United States. They decided in favor of the latter because they believed Baird's system was near the end of its potential development, while the electronic system could be improved.

Electronic Television

Although a number of other scientists became interested in the possibility of television, not all of them approved of Nipkow system. Its main drawback was that the scanning disc had to be rotated mechanically, limiting the speeds that could be achieved and, therefore, the clarity of the picture. The two pioneers of an "alternative" method were A. A. Campbell Swinton, a British scientist, and Boris Rosing, a Russian engineer, working independently. Campbell Swinton concluded that an electronic cathode ray tube, in which the image would be picked up on a thin plate coated with photosensitive substance, should replace the mechanical scanning disc. The plate would then be bombarded with electrons form a "gun" at the other side of the tube.

A German named Karl Braun had invented the cathode ray tube in 1897, but it was Campbell Swinton and Rosing who showed how it could be used for television. Rosing filed a patent in 1907. In 1908, unaware of this, Campbell Swinton wrote a letter to Nature magazine, revealing his proposals. The system described in Nature is, with some refinements, still the basis of television today. What he did not do was to put his principles into practice. This was what John Logie Baird did. But his misfortune was that when he began experimenting with television in 1922, he used the Nipkow disc, even though it had been superseded by the work of Rosing and Campbell Swinton; and as evidence accumulated that he ought to change to electronic scanning, he still obstinately refused to do so.

Early Experimentation

While Baird was transmitting images in London, other people were working along the same lines elsewhere in the world. Denes Von Mihaly, a Hungarian scientist, was also committed to mechanical scanning, but instead of Nipkow disc, he preferred a revolving drum that contained mirrors tilted at different angles. This mirror drum was later adopted by Baird and Alexanderson because of the better quality of the mages it could get. Nevertheless, it still suffered from the essential defects of all mechanical systems. Mihaly, like Baird, Alexanderson and other pioneers, was transmitting 30-line pictures. In Japan, a young lecturer called Kenjiro Takayanagi managed to transmit the Japanese character "I" in 1927. Meanwhile in the Soviet Union, scientists at the Leningrad Polytechnic Institute "were under the impression that they were inventing television". But they pursued electronic television right away, instead of going through a mechanical phase.

In addition, the Americans were "inventing" television, too. Charles Francis Jenkins had started experimenting in 1923 using Nipkow discs. In 1925, shortly after Baird's first show, Jenkins gave a similar demonstration in the United States. Different corporations involved in radio and telephone technology were sponsoring research on television. At General Electric, Ernst Alexanderson was working on mechanical scanning. At Westinghouse, Vladimir Zworykin, who had experimented with television in Russia before emigrating in 1919, was more interested in electronic scanning. David Sarnoff, head of RCA had no doubts as early as 1923 that television was here to stay: "Every broadcast receiver for home use in the future will also be equipped with a television adjunct by which the instrument will make it possible to see as well hear what is going on in the broadcast station."

On April 7, 1927, the American Telephone & Telegraph Company (AT&T) gave a public demonstration of the apparatus that had been developed in its Bell Laboratories. In the course of it, the U. S. Secretary of Commerce, Herbert Hoover, gave a speech, which was broadcast from Washington and watched in New York by an "invited audience of business executives, editors, and bankers." The following day, the event was a lead story on the front page of the New York Times: "Far Off Speakers seen as well as heard here in a test of television." The audience in New York saw two broadcasts: the first, in which Hoover spoke, was transmitted by wire from Washington; the other was sent by radio from AT&T's experimental station in Whippany, New Jersey. The Whippany broadcast featured a comedian called A. Dolan.

Five months later, in Los Angeles, there was an equally historic development: Philo T. Farnsworth produced an electronic television system that worked. What it made it particularly extraordinary was that the inventor himself was only twenty-one years old; a most unlikely figure to be involved in the creation of television. He came from a poor farming family in Idaho. When he was fifteen, and after reading a stack of popular science journals, he found at home, he drew on a blackboard a complete plan for an electronic television system, to the astonishment of his chemistry teacher, who encouraged him to persevere.

After some years of experimentation, Farnsworth felt confident enough to apply for a patent. This was strongly contested by RCA, whose president, David Sarnoff, had decided that television had a future and that RCA, through its subsidiary company NBC, ought to dominate it. Nevertheless, the mighty corporation was unsuccessful, and in August 1930, the 24-year-old Farnsworth won his patent for electronic television. Soon afterwards, Vladimir Zworykin, then working on electronic television for RCA, visited Farnsworth in San Francisco. RCA's lawyers then began the long and expensive process of trying to buy Farnsworth's patents.

Even though they were stuck with mechanical equipment, other American pioneers were making progress. In May 1928, General Electric began making regular thrice-weekly broadcasts from radio station WGY in Schenectady, New York. Ray Kell, who was one of General Electric engineers at the time, says that on Christmas Eve, 1928, he sent a greetings message that was picked up by his parents on a television receiver in Indiana. Most of the broadcasts from Schenectady were less adventurous (only the faces of men talking, laughing, or smoking) but there were a couple of notable breakthroughs. On August 22, 1928, WGY transmitted Al Smith's speech accepting his nomination for the U. S. Presidency. The event took place in Albany, New York, and 24-line pictures were sent back to Schenectady over a telephone wire. In September 1928, WGY broadcast the world's first television drama, The Queen's Messenger, by J. Hartley Manners. The play had only two characters in it, which was just as well, since anything with a large cast would have defeated General Electric's primitive equipment. Each camera could only scan an area twelve inches square, enough for a human head and not much else. Three cameras were used: one for the actress, one for the actor, and one for the two "doubles." These doubles had an essential part to play, as the main actress and actor could not move their heads without going out of focus. Whenever the script called for some other shot—a hand holding a glass of wine, for instance—there was a switch to the third camera, where one of the double's hands would be seen. By the time WGY transmitted its second play, eight months later, the cameras had been improved, and the 12-inch frame was replaced by a stage eight feet wide, eight feet high and six feet deep.

WGY was not the only American television station operating in 1928. From June onwards, Charles F. Jenkins was broadcasting four-line silhouette pictures from Washington Monday, Wednesday, and Friday evenings. By the end of the year there were eighteen experimental television stations licensed in the United States. Progress was spurred by the Americans' knowledge that they were engaged in a race with the British. Research into the new equipment took place in secrecy. NBC's owner, RCA, was doing well: late in 1928, it began secret tests with Zworykin's Iconoscope, a cathode ray tube camera, which differed from the other cameras, both electronic and mechanical, by "storing" an image before scanning it, thus reducing the amount of light that needed to be cast on the image. RCA's first experimental television transmissions began in 1929 by station W2XBS New York in Van Cortlandt Park and then moved to the New Amsterdam Theater Building, transmitting 60 line pictures in the new 2-3 mHz band allocated to television. A 13" Felix the Cat figure made of papier-mâché was placed on a record player turntable and was broadcast using a mechanical scanning disk to a scanning disk receiver. The image received was only two inches tall, and the broadcasts lasted about two hours per day. By 1931, the station became part of NBC and began to transmit from 42nd Street.

After many years of research and development, an all-electronic television system emerged from the laboratory in 1933 for actual field tests. These tests were carried out in Camden, New Jersey, using a video transmitter and connected to it by a coaxial line. Iconoscopes (television cameras) were used to pick up scenes in both the studio and out-of-doors. The use of the iconoscope permitted transmission of detail, outdoor pick-up, and wider areas of coverage in the studio. A scanning pattern of 240 lines made it possible to obtain a picture with good definition, but as the frame frequency was 24 cycles, without interlacing, flicker was quite noticeable. In 1934, the number of lines was increased to 343, and an interlaced pattern having a field frequency of 60 cycles and a repetition rate of 30 frames per second was adopted.

General Electric's Role

The man speaking in the second section of the clip, applauding the success of the recording and explaining its significance, was confirmed to be Ernst F. W. Alexanderson. Alexanderson was born in Uppsala, Sweden, on January 25, 1878. He received a degree in engineering from the Royal Technical University in Stockholm in 1900. He arrived in the U. S. in 1901 and by 1902, he was working at General Electric, first as a draughtsman, but was quickly transferred to Steinmetz's Consulting Engineering Department. He received his first patent in 1904 and eventually became the holder of some 350 patents, the last of which was issued in 1973, when he was 95 years old.

Alexanderson made significant contributions to the advancement of radio and television, in addition to fields such as railway electrification, direct-current power transmission, telephone relays and gun control systems, to name a few. He became the leading authority on the design of high power, high frequency alternators for wireless communications shortly after he was employed by GE. The high frequency alternator he invented enabled Reginald A. Fessenden to transmit the world's first long-range radiobroadcast of both voice and music on Christmas Eve, 1906. He later developed a 200 kW alternator that was used by President Woodrow Wilson to send messages to commanders in Europe. This alternator was completed at the end of the war, when President Wilson delivered the ultimatum to Germany that led to the Armistice. Shortly after this, he sent the first facsimile around the world.

Some of GE's Contributions to Television

GE's experiments with various schemes for practical television led Alexanderson to suggest a method employing multiple spots of light (instead of one that had been used in previous experiments) and a mirror drum. The drum, coupled with a high speed electric motor, projected light onto a screen about 4 × 4 feet. Seven light sources were arranged close together in a star formation. This experimental machine was tested on September 18, 1926. Because each light source would have required an independent control (a problem that was never practically solved) the images produced were quite crude. Overall, the machine had a number of disadvantages, and when R. D. Kell became supervisor of Alexanderson's television laboratory in 1927, he implemented the use of equipment that was already being widely used in the U. S., U. K., Germany, and France. Alexanderson continued make significant contributions to television experiments at GE.

Some major historical moments in television testing at GE include:

- January 13, 1928. Television broadcast demo received by three television sets in Schenectady, in the homes of Alexanderson and two GE executives. Images were said to be as good as those in a lab and detail in faces could be seen.
- August 21, 1928. First outside television broadcast by GE. Listeners

and viewers tuned into WGY or into the shortwave radio stations 2XAF and 2XAD and were able to see and hear Governor A. E. Smith, Democratic candidate for President, make his acceptance speech for the nomination in Albany, NY. By September 1928, WGY was making four weekly television broadcasts of 15 to 30 minutes.

- February 12, 1931. GE reported successfully transmitting images of Professor Karolus to his home Leipzig, Germany. Images were said to be so clear that the professor's eyeglasses could be distinguished. Same report discussed GE's ability to film motion pictures of television images that have quality as good as or better than the television image itself.

OUR FILM: TELEVISION PICTURES

The Demonstration

The man who appears in the television image follows a traditional demonstration pattern that was fairly standard and familiar at this point in time, moving his head from side to side and smoking a cigarette to illustrate the resolution of the image. During the first public demonstration in the U. S. of television broadcasting, on January 13, 1928, a man reportedly smoked a cigarette, and then turned his head from side to side. Similar demonstrations followed. Because this demonstration was recorded on film, it could not exploit the spectacular "liveness" of television, as many previous demonstrations did. It was, however, an attempt at recording the live television transmission, which up to this point seems to have been unsuccessful.

The People

The man speaking on the success of the filming of the television image was easily confirmed to be Dr. Ernst F. W. Alexanderson of the GE Company by consulting numerous websites with photographs of him. Information on the man doing the demonstration in the clip, noted by the Fox Hearst Corporation library index card to be Mr. Lowell J. Hartley, has proved difficult to find. The only mention we have found of him was in a book entitled Alexanderson: Pioneer in American Electrical Engineering. He is briefly noted to have worked extensively in television and radio experiments at GE. In this book, his name is spelled "Hartly." We made a call to the Schenectady Museum, which holds many GE-related materials, including a photographic archive, which unfortunately has not been able to give us any more information.

The Equipment

The mechanical television equipment Kell introduced to GE's television laboratory is similar to what Alexanderson demonstrates in Television

Pictures. What we are seeing in the film is an advanced version of the scanning disc system invented by German Paul Nipkow in 1884, which, as discussed above, was used until the late 1930s by many companies, with photo-electric cells for projection illumination. The cells appear to be Alexanderson's version of Dr. A. Karolus's Kerr cell, developed in 1924. The disc in this clip had 48 apertures, although GE often used 24-aperture discs around this time. This disc then would create 48 lines per image on the television display, resulting in the wavy lines that we see in the film.

The Date

A library index card from the Fox Hearst Corporation lists the date of this film as March 12, 1931, with Bockhorst Haxton as cameraman and a length of 2500 feet. That the footage from the GE laboratories might have been included in a Fox or Hearst newsreel around this time is indicated by a March 23, 1931, article in the New York Times entitled "Newsreel subjects." The article reports that the highlight of the newsreel shown at the Embassy Theatre in New York City that week was a television demonstration from GE's "House of Magic" in Schenectady, NY. The New York Times proclaims that the image "appears to justify the company's contention that if the rate of progress is maintained at its present pace the time may not be far distant when home sets will be practicable." The article also lists the other items shown on the newsreel that week, which include an interview the president of Boston University discussing the individuality of various cities, an interview with the President of Poland, a 4-year-old operatic singer in Mexico. The newsreel described was a combination of Hearst and Fox material; unfortunately, neither Fox nor Hearst has a record of a completed newsreel containing the footage in Television Pictures.

If this is, as Hartley claims, the first time a television image was captured on film (and, as discussed below, we've encountered no evidence to the contrary), we estimate the filming of the television image to have taken place in January or February 1931, based on a New York Times article (as well as a more brief article on the subject in the Chicago Daily Tribune) dated February 12, 1931. The article reports a successful transmission of television images by GE from Schenectady to Leipzig, Germany, and notes the successful creation of filmed motion pictures of television images that would soon "be adaptable for news reel display." In an attempt to pinpoint the date and other details, we contacted a special collections archivist at Union College's Schaffer Library in Schenectady, which has on hand many of Alexanderson's papers and historical materials from the General Electric Company. Unfortunately, the files on Alexanderson do not offer any additional information about the material in Television Pictures.

The Precedent

Hartley says on the clip, "You are now looking at the first motion picture of a television image that has ever been produced." Is this true? Our research suggests this is quite possible. Experiments in filming television had been attempted for a number of years by various companies around the world, though these always seemed to fail. We have found reports of filmed television that always turned out to be faked; that is, they were demonstrations behind the scenes at laboratories, but the image made was staged, not actually a recording of the television image. Since GE was falling behind in the race for practical television technology, they turned to other ways of contributing to the advancement of television. One of these was successfully filming the television image, which as we see, they managed to accomplish. Prior attempts seem to have failed due to inadequate light source.

Our research turned up these records of TV images on film:

- Pathe News, 1927: Pathe News, collaborating with Bell Telephone experts, is the first to show how television operates. It's an apparatus over which calls are made (a telephone with a TV receptor).
- British Pathe, "Seeing by wireless," March 6, 1929: Shows early television transmission techniques. But the actual "televised" image is fake (it's a reconstruction).

Significance

GE was working on the creation of a film recording of television about the same time as they successfully transmitted fairly good quality images from Schenectady to Germany. The fact that these two stories were reported in the same newspaper article hints that there is a relationship between the events. As we learn from listening to Alexanderson's short monologue in the clip, they were looking into the use of film as a means to broadcast events from overseas in theaters shortly after they happened. Events could either be filmed and then broadcast on television or broadcast via television and recorded on film for future distribution. GE's proof that a relationship between film and television was possible would excite the public about the broadcast of world news in theaters in almost the same amount of time it would take to hear them on the radio. By showing this television demonstration in theaters as a news event, the audience would be aware of the potential next step in broadcasting.

Although television was a household word in March of 1931, it was not yet physically in the homes of many people. Experiments were still being carried out in the U. S. and Europe, each inventor and institution hoping to be the first to succeed in creating high-definition, practical television. Because a limited number of people would have been able to witness television demonstrations and broadcasts, it's likely this was the first time audiences in theaters had seen television images and were offered insights into the workings of television. Reaching a mass audience would have given GE an advantage over other U. S. companies attempting to market their systems at the time.

Future Directions for Research

While we were able to locate information about television in this period through books, Web sites, input from archivists solicited via the AMIA listserv, and primary sources such as newspapers, we would like to investigate further the individuals involved in this particular production. We would also like to peruse the archives at the General Electric Company in Schenectady in hopes of discovering additional information about Alexanderson's television research, any extant film footage of this and other experiments, employee records for Lowell J. Hartley, and any paperwork, including a script, that relates to this production.

We would like to find out why there are four copies of the same take (Hartley's television image appears four times and, as far as we can discern, is the same take each time) and why one of these takes has no sound, and we'd like to confirm whether this footage appeared in a newsreel released by either Fox or Hearst. (If it didn't, we'd love to know what was in the newsreel described in the New York Times and released immediately after Television Pictures was filmed at the GE House of Magic!) Given our current restrictions, however, we are content with the information we found: the familiar pattern of Hartley's demonstration, the names of the people involved, the status of television demonstrations and research at this time, details about the mechanical equipment Alexanderson explains, and relative confidence that this is (possibly) the first time a television image was captured on film by a sound Movietone camera.

INTERACTIVE TELEVISION

It appears important at this point to bring forward a working definition of Interactive Television, since we have already presented working definitions for Old Media, Television, New Media and Internet. Obviously this is not going to be a very in-depth analysis as it will appear later in this report.

To begin with, it is important to state that Interactive Television is not just the bringing of interactive features to television; it is far more than that. It is much more about transforming television's linear concept onto a non-linear one, and that should include narratives as well. Clearly that change will not happen from one day to another. Many problems – from which I would point technical and social adaptation – have to be resolved before. It is vital that, from this point on, one bears in mind that having a bigger remote control with loads of buttons to control a television might not be synonym of interactive television, even if (when) the industry considers it so. It might only worth another less presumptuous title, like enhanced or augmented television.

Since it is imperative to define Interactive Television to move on with our analyses, here follows a working definition. On the site Itvnews, Interactive Television is defined as "the meeting of television with new interactive technology. ITV is domestic television with interactive facilities usually facilitated through a 'back channel' and/or an advanced terminal. Equally important, interactive television is content that users and viewers can interact with via the technical system. Interactive television is also a way of empowering viewers to use television in new ways".

As we can see, it mentions technology, domestic television, content that people can interact with. It also points an important change: television will be "used" opposing to "watched". But what is more important is that it leaves open the most crucial part of the whole concept: How is it going to be used? In what "new ways" will people be empowered?

The answers to these questions seem to rely on two factors. The first is what can a starving-for-money industry come up with to engage the consumers. The second and most important is what will the users/ consumers feel the need to/for. Related to this subject, I would like to bring your attention briefly to a recent BBCnews article "Turning on to Interaction". It focuses on a very interesting fact – the booming and later fading of many dot.com's companies and the way the producers of ITV learned various lessons of it. The article argues that the web portals started to offer loads of services "... that dot.coms thought visitors needed. But it became obvious that people preferred to go to sites they already knew, had found themselves, or had been told about by friends. People didn't like being told what to do", and that nowadays, "broadcasters are adapting iTV to what consumers are doing with it rather than forcing them into a narrow range of choices". Focusing on the hardware part of the setop box, the article pointed that this technological adaptation must be scalable.

Nevertheless, on content level, there will be also an ongoing evolution process, as Leos de Vos argues: "When ITV developers would be aware of the contexts of the use of their ITV programmes and services they would develop more insight in the potential interactivity in interactive television settings and this would contribute to the success of their ITV programmes and services". We will obviously come back to these matters later in the text.

As a summary for ITV, I would say, for the moment, that the developments in Interactive Television will have to respect the users' tempo, and not present at once all the interactive possibilities, since many of these will be needs and requests, which will be appearing as a result of the change on the viewing side to the user capability of the device.

THE CONCEPT OF INTERACTIVITY

Later in their work, these authors propose that interactivity should be broken on to four different dimensions:

1. *Spatial Dimension, relating to the "Multi-Lateralness"*: "A primary definition of interactive media is activity in two or more directions. At least two actors and two actions are involved: a supplier or sender transmits signals and a user or receiver returns signals in this way

becoming a sender himself or herself. The number of turns varies and depends upon the number of choices the user can make (for ITV this can be: programmes, additional information, camera angles etc.)".

2. *Time Dimension, relating to "Synchronicity and space of time"*: "AII social and communication scientists agree that the immediate succession of action and reaction reinforces interactivity. Asynchronous communication like in using answering devices or e-mail easily leads to a rupture of interaction, a lesser grip on it or misunderstandings".
3. *Behavioural Dimension, relating to "Controlling the Action"*: "The extent of control of the (inter)action process by people is the most important dimension of interactivity in communication science... The importance of the dimension of control in communication science can be explained by the central role of media characteristics in this discipline. However, control of action is just as well a characteristic of face-to-face communication".
4. *Mental Dimension, relating to "Understanding Action"*: "The level of understanding interactors are able to derive from actions and to locate against a background of experiences and circumstances (context) is the most important difference between face-to-face and mediated communication".

Now that we framed interactivity onto a media context, it is appropriate to bring forward the concept of pacing, as it is of greater importance to the future of the media in general!

THE CONCEPT OF PACING

Pacing can be described as "...the control one can have over the media". Obviously it is a matter of great importance when talking about the media and its use to broadcast commercial messages.

There are basically three pacing categories:

1. Pacing can be made by the sender – "External Pacing";
2. It can be made by the receiver – "Internal Pacing";
3. And it can be made by both – "Mixed Pacing".

It is considered that a television broadcast is external pacing and that a newspaper is internal pacing enabled. Internet and other interactive media are considered to be mixed and internal pacing, depending on the specific situation. Obviously Interactive Television will fall under these two categories, giving an old media characteristics of new media, amongst which internal pacing appears to become the dominant one.

COMMUNICATION

According to the Encyclopaedia Britannica, communication is "the exchange of meanings between individuals through a common system of symbol...". Or, in other words, communication is basically the exchange of information between at least two points or persons.

And Encyclopaedia Britannica adds: "...Explicit definitions and theories of communication were not proposed until 20th-century...". It is about such theories that we are going to centre our attentions in the following paragraphs.

Moreover, Encyclopaedia Britannica states, "advances in science and technology gave rise to the mass-communications media". And as we have seen in the history of advertising, one of the key factors for the population reach it has nowadays were developments in the communication means, enabling a more personal communication to be broaden and done for the masses. These features obviously came out of technology developments, such as the printed industry (newspapers), radio and television.

Communication Feed

Communication Feed relates to the direction of an information flow (e. g. a message). One of the biggest criticisms around the studies on communication models is that many believe that communication is an endless process. Being endless, it cannot be totally represented in a linear way, mostly because a communication process implies that there is a change of the direction to which the feed is heading. For that reason, I decided to split my analyses into two different sections: the linear models and the non-linear models.

MODELS OF COMMUNICATION

Linear Models

It was upon linearity that the first communication studies were developed. In fact, linearity is, in a sense, the basis of all communication, as we will see, even though linear models may not be the most effective way to represent a communication process. On its basis, there are some kind of linear events inherent to a communication process.

GUIDELINES TO INTERACTIVE TELEVISION ADVERTISING

The first requirement is to focus on the vast capabilities that real-time feedback enables and to redefine the use of video in the so-called media-mix.

Apparently the tendencies for the video advert are being grounded on a "on time with the viewer" basis, opposing to that of "timed for the viewer". Maybe, it is under the second they ought to be grounded in an Interactive Television environment, since the viewer has more power to decide when and for how long is the advertiser going to have airtime, so users' time is prime time!

This seems to indicate a changing in the shape of "our" well-known 30s advert, to something more like an infomercial that the viewer requires whenever he/she feels the need for. As such, the once highly paid prime time advertising becomes viewer-available-time or, to be more precise, viewer requiring/demanding information time. It is not about buying advertising space on the prime-time slots, nor it is about where, in the advertising pod, is the advert going to be placed. Not that these two notions are part of the past of television, which they are not. The fact is that, in the past, television was synonym of live-feed information, whereas, in the future, it will tend to become an antonym of it, turning into something more like downloadable feed of information – which actually, in real life terms, means information available upon users demand. In other words: create the contents and establish your business relationships today, so that the benefits appear very strong tomorrow.

Apparently, Interactive Television is set to inverse last decades' advertising tendency (in placed adverts) of shorter lengths and increased frequency to something like "when you want" frequency and "for as long as you want" length, not forgetting the most important "what do you want" selection. Moreover, this tendency will not only become noticed on the duration of the advert itself but also on its shape.

We have already mentioned Seth Godin and Michael Schrage's ideas of interrupting/ intrusive commercial messages and how they do not fit onto a new media advertising environment. The first argues that the future of marketing is all about having the consumer's permission to set down a communication process – e.g. Amazon sends e-mails to its clients to inform them about the latest available titles. In Is Advertising Dead?, Michael Schrage argues that the future adverts will be down to three dominant shapes: invitational, solicited and integral. The first – invitational (pointed earlier) – is already distinguished by some throughout banner advertising on the Internet. In the case of ITV, the user gets some kind of invitation to spare some of his/her time to watch a particular advert.

Most likely the advertiser is going to offer something in exchange, as we have argued already in Internet advertising experiments. The second – solicited – can be found when a user accesses a company's web site, browsing for information or data on the company's products or services. We will analyse it further under Demanded Advertising. The third – integral – we have already discussed under Product Placement Television Advertising, which, as we will see further, will get a big upgrade with the introduction of the ITV back channel. To a certain extent, the first two categories will shift the already described "… imagery and Rock and Roll…" television advert to something more calm and direct, with less redundant and much more specific/ direct information. After all, by accepting the invitation or by soliciting information, the user is already showing a pre-disposition to spare some of his/ her time with that specific company. It is down to that company and those who responsible for its communication to make the users' time worth, otherwise that user will not be coming back.

THE ECONOMIES BEHIND INTERACTIVE TELEVISION

As we have seen along this report, there are many players in the advertising industry: the client (also referred to in this report as advertiser), the advertising agency, the media placement/buying companies and, obviously, the medium and the video production. Moreover, in Interactive Television, there is also the Interactive Television Service Provider (I.T.S.P.) who is responsible for the entire expensive technological infrastructure and, obviously, the interactive contents producers/companies.

Each one of these players undertakes a specific task in the advertising process, from having a communication need to creating, delivering and measuring a communication strategy. Clearly, all of them need, if not eager, to draw the maximum revenue possible out of the process. But, on the end of the line, is the consumer, who is very important on an advertising process. In the new media, it seems that the consumer is the most important character – after all, it is on the consumers' joy that all the players' efforts are or should be focused on. Basically, if the audience does not convert itself in consumers, how can each of the players pay the following player? Ultimately, if the advertiser does not get a favourable Return of Investment (R.O.I.) from its advertising efforts, how can its business subsist, and if there are no clients (advertisers), obviously there is no advertising to be done. In a sense, all the players depend on each other's subsidies to survive and to ensure a workflow.

The I.T.S.P. might be the one with more investment on the upfront, but we must bear in mind that all the others players should, if not have to, invest on its collaborators' new media capabilities and new media cultural education. So, on a second analysis and bringing some points stressed out in this report, we can see that interactive television advertising is not simply an adaptation of the advertising models of television and Internet. Nor it is about adapting both media advert forms to a new media. It is more like "Don't Repackage – Redefine!", the title of an article from Barry Diller. In fact, the author argued in 1995 that each media talks his own language and as such its contents should reflect it. Following his line of thought, I would point that advertising messages, including its business models, ought to follow the same guidelines.

Media Buying and Media Strategy/ Planning for I -TV

Now that we have considered some changes that might take place on the way television advertising is going to evolve once it is taken to an interactive environment, it is time to envision some problems on the way it is going to be bought and sold. Obviously, we can picture here a big mess regarding who is going to get paid for what, as there are too many intervenients.

This is how the problems begin: put it this way, there are loads of channels being broadcasted on ITV, each one has got its own programming. Each one of them is eager to find a way to make money out of something during its airtime. Even if making money means broadcasting advertisers' programmes (bartering) for free. Now the minute those programmes have the ability to shift the viewer's attention somewhere else – e. g. an advertisers' web site – the TV channel is going to want something in exchange from the advertiser. After all, the TV channel will be loosing audience every time one viewer steps out of the broadcast to an advertisers' web site.

Then there is an I.T.S.P. that acts as a carrier for those channels and, therefore, also delivers the adverts. Moreover, it is responsible for assuring a feedback channel to the viewer/user on behalf of the advertiser, when the user decides to move to a one-to-one dialogue with a particular advertiser. Obviously, the I.T.S.P. is going to want to get paid by its crucial services.

Furthermore, there is the advertising agency that worked out some form of displaying its clients' message on a TV channel (Advert, etc.). Besides, there are the media buying companies who obviously will want a piece of the cake as well. Then and very important, there is the advertiser who is not willing to pay more or not willing to pay at all, unless there is some kind of pay-off – lets say a sale! Tricky, isn't it?

Business Models

Under the above described scenario, it appears appropriate to divide Interactive Television advertising on two major categories: the first that relates to Web Buying aspect of the media; and the second that relates to Video Buying.

The idea for this division is that there will be parts of the medium that will be, at least in terms of concept, very alike to those of Internet (Web Buying). These parts relate to choice menus, such as E.P.G. and the P.V.R., not forgetting some programmes that will also run a web site in which the user can search for more information regarding a specific topic. On the other hand, Video Buying relates more to the traditional aspect of television advertising. Nevertheless, we should bear in mind that the video is to have hyperlinkable features, restricting to this any possible resemblances to the "old television".

On top of these two major areas, there are another two: Specialized Content Experts and Real Time Advertising Planning/ Strategy & Buying. Those two are much more related to the real breakthrough of interactivity on television, as we will see. Having said that, it means also that we must still wait to put them to practice.

Web Buying

As we have seen, this relates to the Internet part of interactive television and can be divided onto two different panoramas: the first – let us call it I.T.S.P. Content – regards to the content made available by the Interactive Television Service Provider I.T.S.P. (the equivalent to ISP but for ITV); the second – let us call it Programme Content – relates to the interactive part of a specific programme.

I.T.S.P. Content

This regards to the possible choice menus made available by the I.T.S.P. The more contents or menus the I.T.S.P. develops, the more chances it gets to have ad space bought. This is obviously a very powerful part of ITV advertising, as it relates, in a sense, to "a front door" or to the "hall of entrance" of the house. Simply imagine how many people can be reached here – virtually 100% of the users. We are talking about the E.P.G. and the P.V.R., which are clearly two of the most wanted features of future television.

However, if, on one hand, one could think of it on a more generalist advertising approach, one must not discard the fact that all the users are registered. Therefore, despite the fact that it reaches everyone, it does it by reaching one-at-a-time. As this advertising space can be so targeted, it can and should be mostly used as a teaser, just like banner advertising is used on the Internet. One last detail: the media selling here is done by the I.T.S.P. and clearly all that we talked about in terms of targeting and evaluation for Internet also applies here.

Programme Content

This relates to the content presented in the interactive part of the programme – e.g. banners in the site of the programme, which can be very targeted since they reflects mainly the "real" audience interest on the subject of the programme. Banners' targeting is based on the increasing depth of the users' required information. It seems perfect for documentaries and kids programmes.

Another hypothesis can be that of the site or some parts of it being of a sponsored content kind, offered by an advertiser or group of advertisers, which for the programme producers means a bonus, since they do not pay for nothing and still have a new tool to get closer to their audience. The selling of ad space is made by the TV programme producers and what they sell is space in their site. But again the I.T.S.P. is going to want something.

VIDEO BUYING

Video Buying relates to the video advertising capabilities we already know from the old television advertising practices, although they will be enhanced by the inherent technology of ITV. I think we can split the Video Buying onto two main categories: Programme Buying and Advertising Buying.

Programme Buying

Content sponsorship, product placement and bartering are three practices we have already covered under television advertising. Nevertheless, linking or deep linking the programme to a company's site can enhance all of them. How the link is going to be charged is still to be defined, but one thing is sure: the programme producers are going to want a piece of the cake as well.

Advertising Buying

As we have seen, this relates more to the traditional advertising part of interactive television and can be divided onto two different panoramas: the first – let us call it Live-Feed Advertising – is the television advertising we know nowadays (day and night time based buying), though it will take into account the accurate number of people watching a specific channel at a precise given time; the second – let us call this option Demanded Advertising – is based on a viewer's curiosity or necessity, probably starting with a trip to the advertisers' web site, in which more information is asked, such as a video on a specific product. It is under this panorama that some of the changes in television advertising are likely to became more evident – e. g. Infomercials, interactive content advertising.

Live-Feed Advertising

In live television, maybe this applies strongly to sport or major news events. It is live action, and people usually ache to see live events. In programmed television, just because there is a big fuzz about Interactive Television that doesn't mean all "old" traditional television channels are dead. Although tailor-made TV or even self-programmed TV might weaken pre-made television, there are still many people for whom the interactive concept applied to their television will not be worth a penny – e. g. elderly, sick people, hospitals. And we must not forget that many people will not feel thrilled at all with the idea of adopting a more active approach to the act of watching telly.

Having said that, it is evident that any user will have the option of bypassing the ads, fact that can be seen as a drawback for the advertiser. But I think it is only a drawback if we look at it on an old-fashioned television point of view of "having the maximum possible reach". If we look at it on another perspective, more up-to-date, there is the possibility of knowing precisely how many people watched a given advert. That is very good news for the advertiser.

Another line of though is that of the growth of specific television channels that may survive on an ordinary type of television advertising – just think of the amount of specialised magazines available. And to sponsor them, there are loads of advertisers willing to supply programmes so that the channel has broadcasting content.

Another option is that of running a channel on already available and produced content. There is a lot of television content on archive – e. g. Portugal's SIC GOLD, owned by one of the private networks (SIC) runs 24h over 24h of old programming. So why not make the archives worth some profits? The programmes are already paid. So what might come from advertising are strictly profits. As we have seen, all this kind of Live-Feed Advertising will follow a traditional television advertising orientation.

Nevertheless, we must not forget two factors: the first is that it can be accounted for its real number of viewers, opposing to the estimated number of old television; the second is that it can be also measured for its effectiveness, which on an advertiser point of view is extraordinary.

In terms of media buying, we have once again the television channel, the I.T.S.P. and maybe the programme producers (if it was not produced by the channel) wanting to get paid, and again the advertiser wanting to pay upon results.

Demanded Advertising

Picture those late night telemarketing broadcastings and imagine that someone asks for something like that (horrible vision of course)! Now imagine someone asking a company to "enlighten" his/her knowledge of a particular product or service. If someone is asking specialised information to a company, it means interest or a need, both very important words about which we have talked under Internet advertising. This is where advertisers can and should make the most of the capabilities of the media, creating real interactive solutions to entertain and inform their potential customers, and taking all the necessary time, users' time, to build a relationship of trust and need. This relationship requires the maximum attention paid to the users. After all, the control is in their hands.

SPECIALISED CONTENT EXPERTS

Imagine that you are a fishing enthusiast and you live in a certain remote region of your country. You would like to know your local weather forecast for the next days, true? Nothing new: regional television has been providing this kind of information for years. But what they do not know is your hobbies.

Interactive Television (channels, advertisers and I.T.S.P.) can know them, and so after the weather forecast they can put on specific fishing adverts or shows targeted solely at you. It can even go further in depth! In old television, some adverts were re-adapted (e. g. Renault Clio MTV series). The advert built around the famous James Brown song "GET –UP!!!" was originally produced in Argentina and was broadcasted in two different versions. They were both the same, apart from one of the last camera shots (that was inside the car where some friends shook their heads to the James Brown tune) and the packshot. In the Portuguese broadcasted version, the group of friends was not the same as in the UK version, nor did the packshot used the road-rails as a vanishing point to the product's U.S.P. sentences. Instead, it used a black background with some graphics resembling Vu's meters. After all, the U.S.P. of that particular series was its "Extra-Sound-System".

It should not be impossible to envision adverts with more or less humour, according to the viewers' film preferences. After all, through data-mining it is possible to know what kind of films a certain person usually watches. If it shows that a good percentage are comedies, then it is just a question of broadcasting that specific user the version of the advert with the funny gags. In principle, this is quite targeted tailor-made advertising. Now imagine the concept taken further to ITV programmes and shows, the so-called tailor-made TV. Impressive? Not that much. After all, it is just a question of having a computer program (Software Agent) analysing your daily television viewing data.

Now imagine that the same computer programme, by crossing the weather forecasts in your region – for the area of the river where you usually fish and for an area of another river 100 km away, near the football pitch where your kid's team will be playing – proposes you the following: you leave for fishing tomorrow morning and buy the bait on a specific shop on the way; you go fishing and then have lunch in a quiet restaurant next to the river falls (10 km off the place where you fished), watch the game and come back. Quite a day's programme!

Now imagine this applied to television content, something like your "very smart video recorder" that knows your interests and is able to learn some new everyday. After a while, it would be able to start pointing you interest related programmes and new tendencies for your interests. I would call it a Specialized Content Expert, which knows exactly what pleases its users in terms of content, and therefore seeks to deliver them the most individually and appropriate contents.

Now picture the I.T.S.P. developing the service, maybe together with mobile phone operators network, so that the consumers – upon permission – are advised whenever the agent feels it is relevant to "bug" its owner. How many companies would line up to sponsor such feature?

REAL-TIME ADVERTISING PLANNING/ STRATEGY & BUYING

The sentence itself is a paradox: how can one plan and draw strategies in real time? Obviously if it is a planning with strategic outcomes, it cannot be done in real-time! Nevertheless, lacking a better word combination and for the sake of the arguments to be presented, let us stick to a simpler "Real-Time Advertisement" R.T.A., now that we know what is being done "on the fly" behind the concept that we are going to develop.

We have talked about Specialized Content Expert and its clever use of software agents. Now imagine a further developed version of these agents working and delivering to advertisers results in real time, whose reflections reach the consumers. Let us say you are watching a live TV show with interviews, and that one of the conversations leads to washing machines. By a word recognition system or with an employee inserting key words in a computer, the I.T.S.P. manages to deliver in real-time to a group of potential advertising partners the topic of the conversation. Moreover, it can also provide the exact number of viewers, so that the advertisers can measure their effective reach, which represents the possibilities they have to establish one-to-one dialogues, otherwise also seen as future potential consumers. The advertisers ask the I.T.S.P. to trigger the broadcasting with a link to their site, and that is when you get some kind of invitation from a washing machine producer to go and visit its site, where you can find out more about its washing machines… These invitations would have to be very simple – like a simple text overlaying the video – so that they could be on in few seconds. The point is: if an invitation proper targeted is on at the right time, maybe it can "push" some people into the site and give the advertisers a bit of time with the consumer.

Let us say now that you are browsing the Web on your telly, looking for mobile phones. Imagine that you get the same kind of message or a pop-up "self-close" timed window, calling you to turn to television mode because on channel 507 there are some guys chatting about mobiles, or that anything else warns you that the advert for the new Nokia phone is going to be on in 5's in channel 57. Visibly on these two proposals there is the privacy issue, but this mode could be turned ON or OFF by the user, which means that permission was granted or not. Or imagine again some kind of reward for having the service ON, like a free film on V.D.O. at your choice.

As it looks, Media Placement can almost be "Real Time" and practically free. The open contact between the I.T.S.P. and advertiser partners is cheap. What represents some investment is software development, but then again, as it is an evolution version of a software agent, its costs are not exorbitant.

5

Test Materials for Subjective Assessment of Stereoscopic Television Pictures

ASSESSMENT FACTORS

Assessment factors generally applied to monoscopic television pictures, such as resolution, colour rendition, motion portrayal, overall quality, sharpness, depth, etc., could be applied to stereoscopic television systems. In addition, there would be many factors peculiar to stereoscopic television systems.

Some of them are listed below, and further studies are required to identify others and to establish physical definitions:

- *Depth resolution*: Spatial resolution in depth direction. Coarse resolution in depth direction may reduce picture quality in stereoscopic television.
- *Depth motion*: A factor related to whether motion or movement along depth direction is reproduced smoothly.
- *Puppet theatre effect*: This describes one type of distortion in reproduced 3-D images. Stereoscopic objects are sometimes perceived as unnaturally large or small.
- *Cardboard effect*: This describes another type of distortion in reproduced 3-D images. The 3-D positions of stereoscopic objects are perceived stereoscopically but they appear unnaturally thin.

ASSESSMENT METHODS

The methods described in Recommendation ITU-R BT.500 could be applied for the evaluation of the general picture quality of stereoscopic systems as well as sharpness and dept. When a reference image is available, double-stimulus continuous quality-scale or double-stimulus impairment scale methods can be used. Examples include comparison of display systems, quality assessment of coding systems, and so on. When no reference is available, the categorical judgement method can be used, for example, to identify the merits of stereoscopic systems. Evaluation methods for the assessment of particular factors of stereoscopic television systems require further study.

Viewing Conditions

Two major factors peculiar to stereoscopic display should be taken into consideration, namely the display frame effect and inconsistency between accommodation and convergence.

Stereoscopic pictures appear highly unnatural when objects positioned in front of the screen approach the screen frame. This unnatural effect is called "the frame effect". The effect is generally reduced with a larger screen, because observers are less conscious of the existence of the frame when the screen is larger.

The human eye focuses on an object according to the distance to that object. At the same time, we also control the convergence point (gaze point) on the object. Therefore, there is no inconsistency between accommodation and convergence in our everyday life. However when viewing stereoscopic images, the focus point (accommodation) must always be fixed on the screen, independent of the convergence point which is derived from the disparity of the signals. Otherwise, the observer cannot focus clearly. Thus, an inconsistency between accommodation and convergence is introduced in stereoscopic systems.

It is generally said that the minimum value for depth of field of the human eye is ± 0.3 D (Diopter: reciprocal value of distance (m)) [Hiruma and Fukuda, 1990]. This means that we can perceive the image without defocusing when the object is located within ± 0.3 D. When viewing stereoscopic television, the accommodation point is fixed on the screen, and therefore stereoscopic pictures should preferably be displayed within this range. Since ordinary television programmes include images at infinite distance (that is D ? 0), the desirable range of depth to be displayed with stereoscopic systems is considered to be within 0 to 0.6 D. Therefore, 0.3 D, i.e. 3.3 m, is considered to be the optimum viewing distance.

Camera parameters (camera separation, camera convergence angle, focal length of lens), resolution of the system and the frame effect should be taken into account in determining viewing conditions (screen size). In the case of HDTV when watching at the standard viewing distance of 3 H (H denotes picture height), the viewing distance of 3.3 m corresponds to a 90-inch screen. In the case of standard definition television (SDTV) when watching at the standard viewing distance of 6 H, this distance corresponds to a 36 inch screen. A subjective assessment of the relationship between screen size and depth perception was carried out with stereoscopic HDTV system, and the results showed that the most natural depth perception was obtained with a screen size of 120 inches, which corresponds to viewing distance of 2.2 H.

Test Materials

Test materials for screening observers and still and motion sequences of natural scenes. The 3-D effects obtained from stereoscopic pictures depend largely on the shooting conditions, such as camera separation, camera convergence angle and focal length of the lens. The motion sequences were shot under the condition of camera separation of 65 mm, corresponding to average eye separation, and most of them were produced with the uncrossed camera layout, which gives an orthostereoscopic condition.

TEST MATERIALS OF STEREOSCOPIC TELEVISION PICTURES

VISION TEST

The test charts for the vision test. These 12 tests are selected according to the hierarchy of the human visual system from lower to higher levels. Eight main vision tests (VTs) are described below, and the other four are for the clinical test. Observers must have normal stereopsis, meaning that they must pass test VT-04 for fine stereopsis and test VT 07 for dynamic stereopsis. The remaining six tests are for more detailed characterization. The test charts should be viewed from three times the height of the display screen (3 H). Below, right and left thumbnail images are put side by side for crossed free fusion for explanatory purposes.

VT-01: Simultaneous Perception (Lion Test)

Tests the ability to perceive dichoptically presented images simultaneously and in the correct position. A cage image is presented to one eye and a lion image to the other eye, with its position moving by 12?/s. The size of each image is fixed at 10? so that the observers can capture the images within their paramacula. Observers with normal vision can see the lion in the cage at a certain time within the presentation period.

VT-02: Binocular Fusion (Worth 4-dot Test)

Tests the ability to perceive two dichoptic images in left and right eyes as one image. The image for one eye has two dots, and the image for the other eye has three dots, with one dot in common. Observers with normal vision can see four dots.

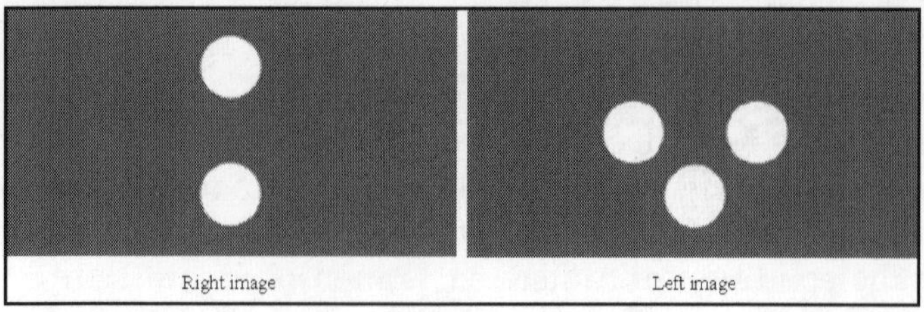

VT-03: Coarse Stereopsis (Dragonfly Test)

Tests the ability to perceive dichoptically presented images with a parallax as one image with a coarse depth. The images for the two eyes are a stereopair of images of a dragonfly with its wings spreading. Observers with normal vision can perceive the wings in front of the display screen.

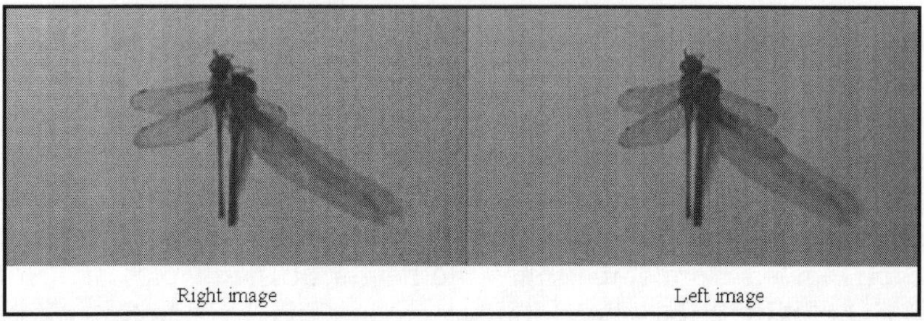

VT-04: Fine Stereopsis (Circle Test)

Tests the ability to perceive dichoptically presented images with a parallax as one image with a fine depth. Nine test lozenge patches are provided and each of them has four circles in which only one circle has a small parallax. Observers with normal vision can perceive the circle with a small parallax in front of the display screen. Table shows the test number, correct answers, and angle of stereopsis at 3 H.

Table. Correct answers and parallax

Test No.	Correct Answers	Angle of Stereopsis at 3 H (")
1	Bottom	480
2	Left	420
3	Bottom	360
4	Top	300
5	Top	240
6	Left	180
7	Right	120
8	Left	60
9	–	0

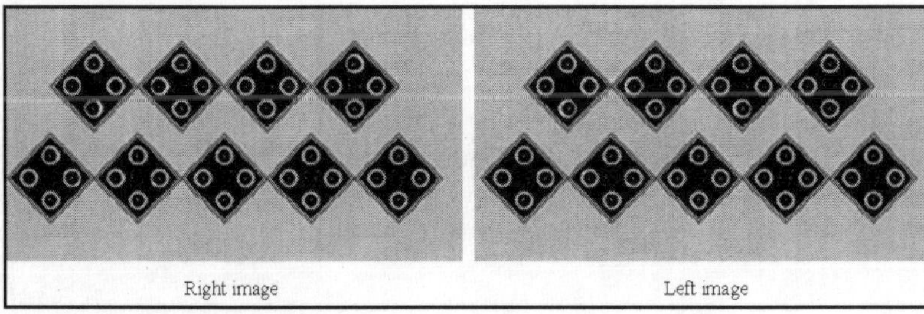

VT-05: Crossed Fusional Limit (Bar Test)

Tests the ability to perceive dichoptically presented images with crossed disparities as one image. A stereopair of bars is presented with its parallax changing by 10?/s.

The fusional limits for the ascending and the descending series can be measured. Observers are instructed to report their fusional break as soon as they perceive double images in the ascending series, and their recovery of fusion as soon as they perceive the dichoptic images as a single image in the descending series.

VT-06: Uncrossed Fusional Limit (Bar Test)

Tests the ability to perceive dichoptically presented images with uncrossed disparities as one image. Presented images are the same as in the crossed case above, but right and left images are swapped.

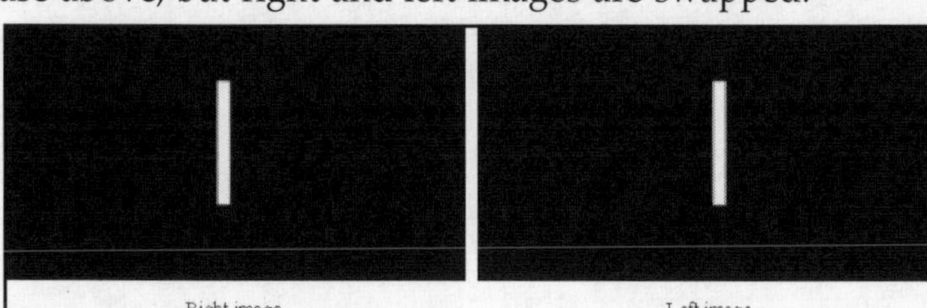

VT-07: Dynamic Stereopsis (Dynamic Random Dot Stereogram Test)

Tests the ability to perceive depth in moving random dot stereogram images. Observers with normal vision can perceive a rectangular shape and a sinsoidal depth motion in the dynamic random dot stereogram.

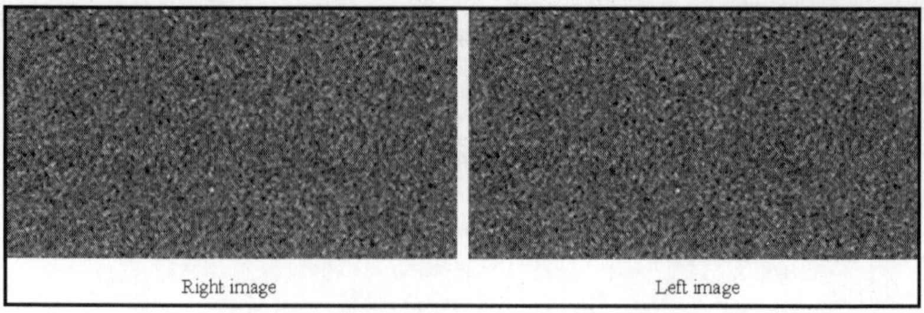

VT-08: Binocular Acuity (Acuity Test)

Tests the binocular acuity with binocular fusion, including any imbalance of monocular acuity which might prevent good stereopsis. The images have four columns and five lines which consists of E characters with a variety of orientation and size. The centre two columns can be seen with both eyes; the left two columns can be seen only with the left eye; and the right two columns can be seen only with the right eye. Observers with normal vision can tell the orientation of the E characters correctly. The character sizes correspond to acuities of about 1.0, 0.5, 0.33, 0.25, and 0.125 at 3 H.

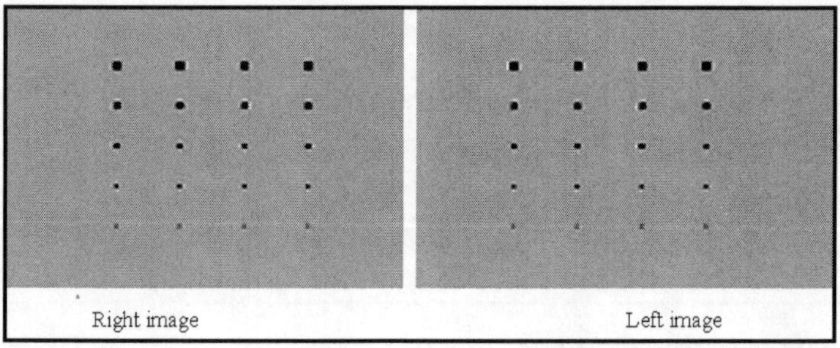

NATURAL IMAGES

The natural images consist of 15 still pictures and 15 motion sequences, as listed in Tables 3 and 4. Some of them are illustrated in Appendix 1. Each image is printed from left to right as left image, right image, left image: the 3-D image can be obtained by fusing the leftmost pair (eyes uncrossed) or the rightmost pair (eyes crossed).

STATUS OF USAGE OF THE STEREOSCOPIC TEST MATERIAL

Use of the test material is restricted to the following purposes only:

- Technical evaluation, including:
  - Research and development of equipment and systems,
  - Testing of equipment in development and production process,
  - Testing of transmission condition for broadcasting and telecommunication,
  - Maintenance of equipment.
- Demonstration, including:

- Presentation at technical conference and workshop,
- Presentation of performance and functionality of equipment, excluding commercial promotion.

Examples of Natural Motion Sequences

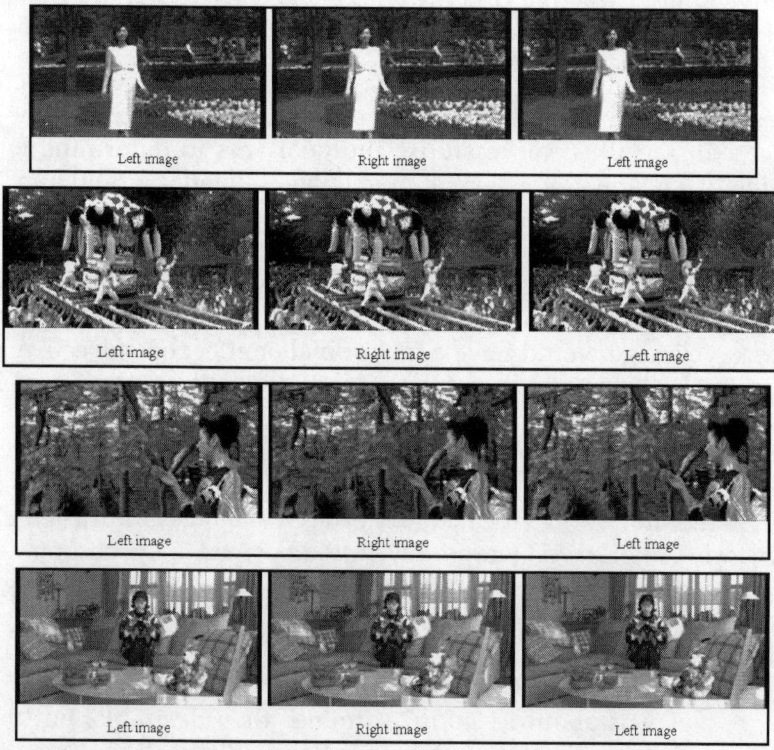

EXPERIMENTAL RESULTS USING THE DOUBLE-STIMULUS CONTINUOUS QUALITY SCALE (DSCQS) METHOD

This Annex documents the application of the DSCQS method to the subjective evaluation of such pictures. The DSCQS method has been used widely and successfully to evaluate the subjective image-quality of monoscopic pictures. Experiences have shown that this method is equally well-suited to evaluating stereoscopic pictures, and that it can easily be adapted to measure other image attributes in addition to image-quality, such as perceived sharpness and depth.

Measuring Perceived Sharpness and Depth with the DSCQS Method

The DSCQS method was easily adapted to measure attributes other than subjective image-quality. This was done by making specific changes to the instructions that were given to viewers. For example, the method was adapted to measure perceived sharpness, and the overall impression of depth of stereoscopic image sequences. Only one attribute (either subjective image-quality, perceived sharpness or perceived depth) was measured in any given testing session.

Illustrative Study Using the DSCQS Method with Stereoscopic Images

The DSCQS method was used to measure the subjective quality, perceived sharpness, and overall depth of a set of stereoscopic and monoscopic image sequences. In the illustrative study, the goal was to determine whether processing one channel of a stereoscopic image sequence would affect these attributes. To this end, the right-eye view of the stereoscopic image sequences was low-pass filtered at three levels: unfiltered, half-resolution and quarter-resolution. In the monoscopic conditions, both eyes saw the filtered view. A survey of the scientific literature led us to expect that filtering one channel of a stereoscopic image would have a much smaller effect on subjective ratings than filtering both channels – subjective rating would be dominated by the unfiltered channel.

In the set-up, the left and right image pairs of a stereoscopic sequence were displayed at 120 Hz using a time-sequential method, on a 29 inch direct-view video monitor or on a 65 inch rear projector. Images for the left and the right eyes were interleaved temporally, and displayed in synchrony with the opening and closing of a pair of Crystal Eyes liquid-crystal shutter glasses, manufactured by StereoGraphics. The liquid-crystal glasses had a light transmittance of about 30%, and response times of 0.2 ms and 2.8 ms for closing and opening, respectively. This meant that the peak luminance entering the observers' eyes was about 21 cd/m2, dimmer than desirable, but the best achievable with the available time sequential display technology. Viewing distance was 4 H. A gray field of 10 cd/m2 was displayed between presentations of the A and the B sequences. It should be noted, that any other method of displaying stereoscopic images could readily be substituted for the time-sequential method without affecting the use of the DSCQS method.

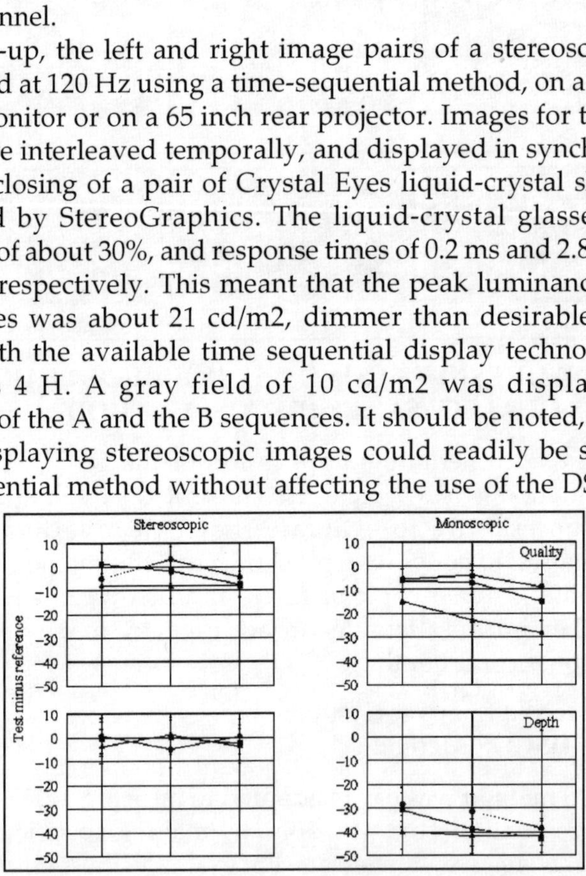

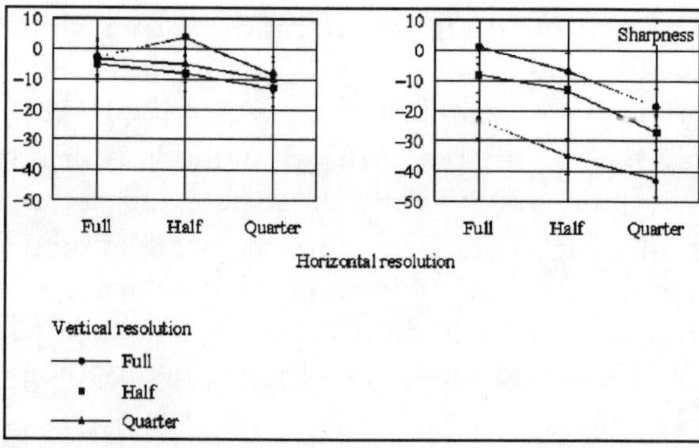

Results of the experiment are presented in Fig. The Y-axis indicates the test minus reference score. A score of zero means that the test sequence was rated equivalently to the unprocessed stereo reference sequence. A negative score means that the test sequence was rated lower than the reference sequence.

Effects of low-pass filtering are evident in the slope and vertical offset of the lines. As expected, in monoscopic conditions, low-pass filtering had a large effect on rated sharpness and image quality. The low ratings of depth occurred because only monocular cues to depth were available in the monoscopic test sequences. Also as expected, in the stereo conditions, all three dimensions (quality, depth and sharpness) were rated higher than in the monoscopic conditions. Low-pass filtering of one channel of a stereo pair caused negligible effects on perceived depth, and minor effects on perceived sharpness and overall quality. Evidently, the high spatial frequency information in the unfiltered left eye image compensated for the lack of this information in the right eye image. The illustrative study, and other work on stereoscopic image sequences with the DSCQS method leads to a conclusion that this method is a valuable and useful tool for studying stereoscopic pictures. The method could be adapted to measure other aspects of stereoscopic image sequences such as presence, power, and naturalness.

SUBJECTIVE METHODS FOR THE ASSESSMENT OF STEREOSCOPIC THREE-DIMENSIONAL TELEVISION (3DTV) SYSTEMS

SCOPE

This Recommendation provides methodologies for the assessment of stereoscopic television systems including general test methods, the grading scales and the viewing conditions.

The ITU Radiocommunication Assembly, considering:

- That a large amount of information has been collected about the methods used in various laboratories for the assessment of critical performance characteristics of three-dimensional television (3DTV) systems;

- That examination of these methods shows that there exists a considerable measure of agreement between the different laboratories about a number of aspects of the tests;
- That the adoption of standardized methods is of importance in the exchange of information between various laboratories;
- That the introduction of three-dimensional television (3DTV) services might require the development of new image formats, image processing, and transmission techniques whose performance will need to be evaluated though subjective methodologies;

Recommends

The general methods of test, the grading scales and the viewing conditions for the assessment of picture quality, described in the following Annex should be used for laboratory experiments and whenever possible for operational assessments;

Assessment (Perceptual) Dimensions

Stereoscopic three-dimensional television exploits the characteristics of the human binocular visual system by recreating the conditions that bring about the perception of the relative depth of objects in the visual scene. The main requirement of current stereoscopic imaging is the capture of at least two views of the same scene from two horizontally aligned cameras. The images of the objects depicted in the scene will have different relative positions in the left- and right-view.

This difference in relative positions in the two views is typically called parallax or disparity, and it usually expressed in pixels or physical distances (a percentage of screen width). It should be noted that parallax should be confused with angular (retinal) disparity. In fact, the same parallax information would produce different angular (retinal) disparities with different viewing distances. The magnitude and direction of the perception of depth is based on the magnitude and direction of the retinal disparities elicited by the stereoscopic image.

Assessment factors generally applied to monoscopic television pictures, such as resolution, colour rendition, motion portrayal, overall quality, sharpness, etc., could be applied to stereoscopic television systems. In addition, there would be many factors peculiar to stereoscopic television systems. These might include factors such as depth resolution, which is the spatial resolution in depth direction, depth motion, that is, whether motion or movement along depth direction is reproduced smoothly and spatial distortions. Two well known examples of the latter are the "puppet theatre effect, i.e., when objects are perceived as unnaturally large or small, and the cardboard effect, i.e., when objects are perceived are perceived stereoscopically but they appear unnaturally thin.

We can identify three basic perceptual dimensions which collectively affect the quality of experience provided by a stereoscopic system: picture quality, depth quality, and visual comfort. Some researchers have argued that the psychological impact of stereoscopic imaging technologies could be better measured in terms of more general concepts such as naturalness and sense of presence.

Primary Perceptual Dimensions

Picture quality refers the perceived quality of the picture provided by the system. This is a main determinant of the performance of any video system. Picture quality is mainly affected by technical parameters and errors introduced by, for example, encoding and/or transmission processes.

Depth quality refers to the ability of the system to deliver an enhanced sensation of depth. The presence of monocular cues, such as linear perspective, blur, gradients, etc, conveys some sensation of depth even in standard 2D images. However, stereoscopic 3D images contain also disparity information which provides additional depth information and thus an enhanced sense of depth as compared to 2D.

Visual (dis)comfort refers to the subjective sensation of (dis)comfort that can be associated with the viewing of stereoscopic images. Improperly captured or improperly displayed stereoscopic images could be a serious source of discomfort.

Additional Perceptual Dimensions

Naturalness refers to the perception of the stereoscopic image as being a truthful representation of reality (i.e., perceptual realism),. The stereoscopic image may present different types of distortion which make it less natural. For example, stereoscopic objects are sometimes perceived as unnaturally large or small (puppet theatre effect), or they appear unnaturally thin (cardboard effect).

Sense of presence refers to "the subjective experience of being in one place or environment even when one is situated in another". This recommendation presents information regarding methods and procedures for the assessment of the three primary dimensions: picture quality, depth quality, and visual comfort, outlined above. Methodologies for the assessment of naturalness and sense of presence will be added at a later stage.

GENERAL VIEWING CONDITIONS

The viewing conditions (including screen luminance, contrast, background illumination, viewing distance etc) should generally match those used for 2D as described in Rec. BT. "General viewing conditions for subjective assessment of quality of television pictures". The reason why we would like to keep (as much as possible) the same parameters is twofold. First, in practice users will watch 3DTV with the same displays and viewing conditions as 2D. Secondly, the performance of 3DTV will often need to be measured in relation (i.e., "as compared") to standard HDTV. In such cases, it might not be possible to keep all parameters the same. Which one should be changed, if any, should be matter of investigation, and the purpose of the experiment.

Rec. BT. [GVC] specifies two possible criteria for the selection of the viewing distance. The Design Viewing Distance (DVD) is to be selected. The DVD for a digital system is the distance at which two adjacent pixels subtend an angle of 1 arc-min at the viewer's eye; and the horizontal design viewing angle as the angle under which an image is seen at its optimal viewing distance.

For example, when expressed in multiples of the picture's height, the DVD for the 1280xx720 (Rec. ITU-R BT.1543) image resolution system is 4.8H (static images...should be 4.7!); and that for the 1920x1080 family (Rec. ITU-R BT.709) HDTV image resolution system is 3.1H (static images).

For illustrative purposes, Table I reports the design viewing distance in meters for a representative sample of TV set diagonal sizes.

Table. Design Viewing Distance in Meters for Various TV Set Diagonal Sizes

	1920 ■ 1080 Image System	1280 ■ 720 Image System
Diagonal Size (inches)	Design Viewing Distance (Meters)	Design Viewing Distance (Meters)
32	1.2	1.9
42	1.6	2.5
52	2.0	3.1
62	2.4	3.7
72	2.8	4.3
82	3.2	4.9
92	3.6	5.5
102	3.9	6.1

It should be noted since two adjacent pixels subtend an angle of 1 arc-min at the viewer's eye, then at design viewing distance the smallest angular (retinal) disparity that can be represented by the system (i.e., depth resolution) is equal to 1 arc-min (or equivalently 60 arc-sec). This value is about twice the human disparity threshold, which is about 30 arc-sec. Therefore, most viewers should have no difficulty resolving the smallest disparity represented by the 3D system. (This is true for all systems in Table I when presented at the design viewing distance.

Test Material

The selection of the test material should be motivated by the experimental question addressed in the study: e.g., the content of the test sequences (sport, drama, film, etc) and their spatiotemporal characteristics should be representative of the programmes delivered by the service under study). In addition, the selected stereoscopic test sequences content should also be normally comfortable to watch. The visual comfort of stereoscopic images depends critically upon the disparity contained in the image. Accordingly, care should be taken to ensure that the disparity does not exceed the limits outlined in the following section, unless the study is specifically aimed at measuring visual comfort. Moreover, whenever possible the statistics: mean, standard deviation, and range (min/max), of the disparity (screen parallax in pixel) distribution of the test sequences should be measured and reported.

Visual Comfort Limits

Excessive disparity/parallax causes visual discomfort possibly because it worsens the conflict between accommodation and vergence. Therefore, it has been suggested that to minimize the accommodation-vergence conflict, the disparities in the stereoscopic image should be small enough so that the perceived depths of objects fall within a "comfort zone". Several limits have been proposed.

One approach uses a measure of the screen parallax, expressed as a percentage of the horizontal screen size, to specify the limits of comfortable viewing. Values of 1% for crossed/negative disparities and 2% for uncrossed/positive disparities (for a total value of about 3%) have been suggested. According to another approach, the comfort zone is delimited by the depth of field of the eye. For the viewing conditions typical of television broadcast, researchers have assumed a depth of field between ± 0.2D (diopters) and ± 0.3D (diopters). For a 1920 × 1080 (Rec. ITU-R BT.709) HDTV image resolution system watched from the design viewing distance of 3.1H, these values correspond approximately to ± 2% and ± 3% of screen parallax.

Recall that at the design viewing distance two adjacent pixels subtend an angle of 1 arc-min at the viewer's eye. Thus, 60 pixels correspond to 1 degree of visual angle. This allows us to easily specify the comfort limits in terms of retinal disparity (for an average viewer). For example, for 1920x1080 (Rec. ITU-R BT.709) HDTV image resolution systems, 1% (~19.2 pixels) corresponds approximately to 20 arc-min, 2% to ~40 arc-min and 3% to ~60 arc-min.

It should be noted that even though at the design viewing distance two adjacent pixels always subtend an angle of 1 arc-min, the physical separation (e.g., in mm) between those pixels increases with larger displays (the number of pixels remains the same, but the physical size of the screen increases). Therefore, the higher limits (e.g., ± 3%) could result in larger displays in a physical distance between corresponding points (i.e., the parallax of the two views in mm) that exceed the interpupillary distance of the average viewer (~63-65 mm). This could result in increasing discomfort.

In general, since studies using stereoscopic test sequences could elicit some degree of visual discomfort, it is recommended to use, whenever possible, test material whose disparity does not exceed the lower limits, albeit occasional excursions above these limits might be allowed.

Experimental Apparatus

The experimental apparatus (video server, display, etc) should be capable of displaying full resolution HD test sequences, for example using an HDMI frame-packing format (unfortunately for now this is only possible for 720p target!). This would allow greater flexibility in the range of studies that be carried out. Figure below shows an example of how it would be possible to test different 3DTV transmission formats using a simple single stimulus methodology.

As of today, there is no reference display for 3DTV assessment. The display should exhibit very low cross-talk (ideally below human threshold) and capable of receiving a variety of input formats (without having to manually change settings!).

OBSERVERS

Sample Size

Sample size considerations for 3D studies are not different from those for 2D studies.

Screening

Observers should be screened for visual acuity, color, and stereoscopic vision. The latter could be assessed using clinical tests, such as Randot, Titmus, or Frisby stereo tests. These clinical tests usually measures retinal disparities from 20 to 400 arc-sec.

INSTRUCTION TO OBSERVERS

Instruction should be tailored to dimension (e.g., depth quality, comfort, etc) under investigation. Furthermore, ethical guidelines are more stringent than those typically used in image quality assessment since participants might experience visual discomfort. In general, these studies require more care in informing the participant of the motivations of the study as well as any possible negative resulting from exposure to the stimuli used in the study.

Session Duration

If the viewing material is deemed comfortable, then the session duration might be as long as that used for 2D studies (i.e., ~20-40 minutes intermixed with breaks). If the material is known to contain excessive parallax, and thus known to be potentially uncomfortable, then the duration should be limited.

Subjective Methodologies

Many of the standard methods outlined in Rec. BT. 500 could be used, occasionally in a slightly modified form (e.g., different scales), for the assessment of stereoscopic systems. A few methods that could be used for the assessment of picture quality, depth quality, and visual comfort are presented in Tables IIa, IIb, and IIc, respectively.

Use of Reference Test Sequences

All the methods listed in Section 8 should include a "reference" sequence, whenever available, as part of the test sequences set. The "reference" is usually a version of the test sequence that has not undergone any processing (i.e., the original source sequence). For the 3DTV studies, the main "reference" is the original unprocessed stereoscopic sequence. The experimental plan might include also the monoscopic version of the "reference" (i.e., only one view of the original source sequence); for example in visual comfort studies it might be useful to use the visual comfort of the monoscopic reference as the baseline.

Statistical Analysis and Viewers' Rejection Criteria

The statistical analyses and the viewers' rejection criteria should be the same as for 2D studies.

A BEGINNER'S GUIDE TO SHOOTING STEREOSCOPIC 3D

3D is back in style again and it seems like everyone, from Hollywood producers to wedding videographers, is interested in producting stereoscopic 3D content. So how can you get involved by shooting your own 3D content? It's actually quite easy to get started and learn the basics of stereoscopic 3D photography. You won't be able to sell yourself as a stereographer after reading this beginner's guide (it literally takes years to learn all the aspects of shooting and build the necessary experience to shoot good stereoscopic 3D) but I guarantee you will have some fun and impress your friends.

The basic principle behind shooting stereoscopic 3D is to capture and then present two slightly different points of view and let the viewer's own visual system determine stereoscopic depth. It sounds simple enough but the first thing any budding stereographer should learn is some basic stereoscopic terminology. These few terms may seem daunting at first but they will form the basis of your stereoscopic knowledge.

Terminology

Stereoscopic 3D a.k.a. "Stereo3D," "S-3D," or "S3D"

"3D" means different things to different people. In the world of visual effects it primarily refers to CGI modeling. This is why stereographers refer to the craft specifically as "stereoscopic 3D" or simply "S3D" to differentiate it from 3D CGI.

Interaxial (a.k.a. "Stereo Base") & Interocular (a.k.a. "i.o.") separation

Fig. Interaxial Separation between Lenses

The interocular separation (or interpupulary distance) technically refers to the distance between the centers of the human eyes. This distance is typically accepted to be an average of 65mm (roughly 2.5 inches) for a male adult. Interaxial separation is the distance between the centers of two camera lenses (specifically the entrance pupils.) The human interocular separation is an important constant stereographers use to make calculations for interaxial separation. Beware that Interaxial separation is often incorrectly referred to as "Interocular" and vise-versa. In the professional world of stereoscopic cinema it has become the norm to refer to interaxial separation as "i.o." even though it is the incorrect term.

Binocular Vision, Retinal Disparity and Parallax

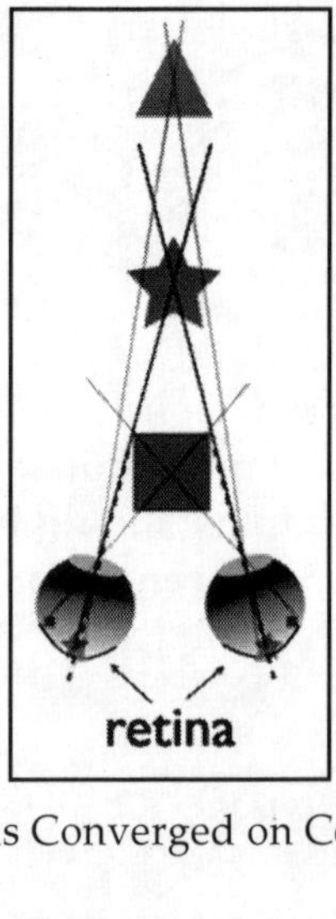

Fig. Eyeballs Converged on Center Object

Binocular Vision simply means that two eyes are used in the vision system. Binocular Vision is very important to most mammals (including humans) because it allows us to perceive depth at close range. Try this: Hold one finger next to your ear. Now stretch your other arm out straight and hold up another finger.

Now bring your two fingers together and touch the tips together. Is was easy right? Now repeat the same procedure but close one eye. Were you able to touch your fingers together on the first try? Now you know how important binocular vision is at close range.

When we look around at objects at different distances from us the images of those objects will be projected on our retinas in slightly different locations for each eye. Our brain can interpret this "Retinal Disparity" and help us determine depth. When we shoot 3D with two cameras from slightly different positions the same thing happens; each camera's sensor registers the objects in the scene in slightly different horizontal positions. We call this difference "parallax."

Convergence & Divergence

Binocular Vision and Parallax are the primary visual tools animals use to perceive depth at close range. The wider an animal's eyes are apart (its interocular distance) the deeper its binocular depth perception or "depth range."

At greater distances we start to use monocular depth cues like perspective, relative size, occlusion, shadows and relation to horizon to perceive how far away objects are from us. Of course it would be difficult to look at double images all day so instead our eyes naturally angle in towards the object of interest to make it a single image. This is called convergence.

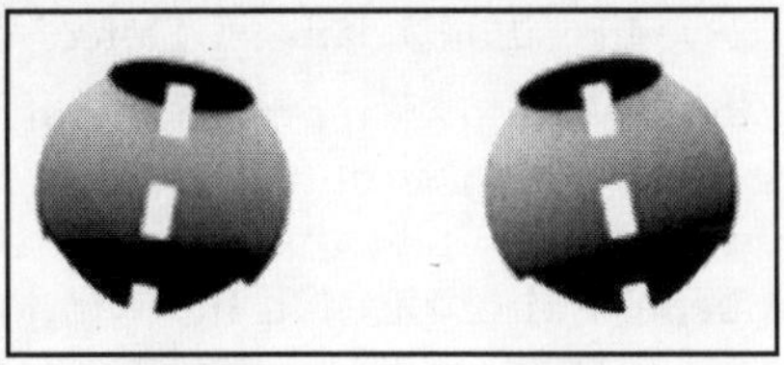

Fig. Converged Eyes

Here's an example of how your eyes use convergence in the real world. Hold a pen about one foot in front of your face and look directly at it. You will feel your eyes both angle towards the pen in order to converge on it, creating a single image of the pen.

What you may not immediately perceive is that everything behind the pen appears as a double image (diverged.) Now look at the background behind the pen and your pen will suddenly appear as two pens because your eyes are no longer converged on it. This "double-image" is retinal disparity at work and it is helping your brain determine which object is in front of the other.

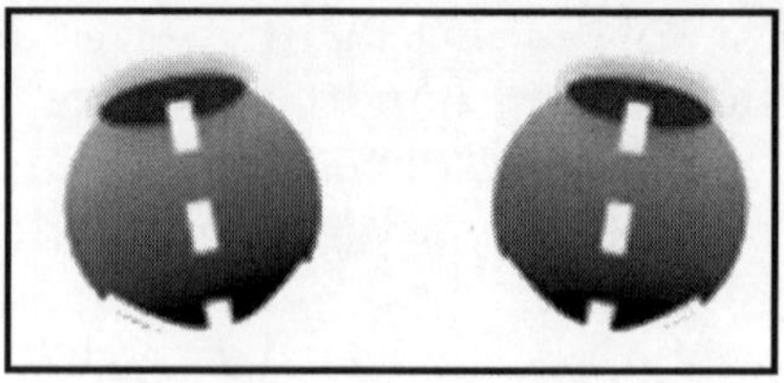

Fig. Diverged Eyes

What never happens to your eyes in the natural world is divergence, which would mean that your eyes would angle outward. This is because the furthest point you could possible attempt to look at is at infinity and even infinity would only require that your eyes be angled perfectly parallel to each other. This is why stereographers should avoid background parallax values in their scene that may require the eyes to diverge when viewed. This is easy to keep in check through some simple math but we will cover that a little later.

Stereo Window, the Screen Plane and Negative, Zero or Positive Parallax

Fig. Perceived Position Relative to Stereo Window

Simply put, the "Stereo Window" refers to the physical display surface. You will be able to visualize the concept if you think of your TV screen as a real window that allows you to view the outside world. Objects in your stereoscopic scene can bebehind or outside the window (positive parallax,) on the window (the Screen Planeor zero parallax,) or inside, between you and the window (negative parallax.)

In the same way objects appear in different horizontally offset locations on our retina to create parallax separation, stereoscopically recorded and displayed objects will appear to have different horizontal offsets (parallax) depending on their depth in the scene. If an object has no perceivable amount of parallax then we consider it to appear on the screen surface just as the star in the illustration. This is why converging on an object will make it appear to be at the screen. This can be done by converging the cameras on the objects while shooting, or by sliding the images horizontally in opposite directions during post production.

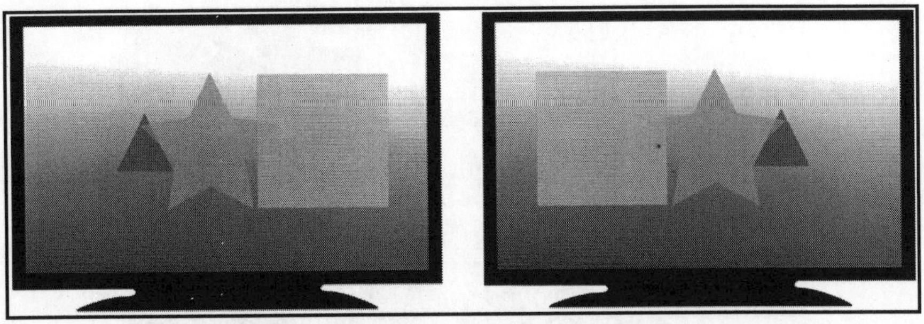

Fig. Left Eye Presented View Cersus Right Eye Presented View (Exaggerated Separation for Demonstration Only)

If an object's left image is to the left of the corresponding right image then that object has positive parallax and will appear to be behind the screen. If an objects left image is to the right of the right image then it has negative parallax and will cause your eyes to cross, which will suggest to your brain that the object is in front of the screen. This is the basic principle behind stereoscopic shooting and emulating human binocular vision with two cameras.

Respecting the Stereo Window

We discussed briefly how the display screen represents a window and objects can be behind, at or in front of the window. If you want an object to appear in front of the window it cannot touch the left or right edge of the frame.

If it does the viewer's brain won't understand how the parallax is suggesting the object is in front of the screen, but at the same time it is being occluded by the edge of the screen. When this contradiction happens it is referred to as a window violation and it should be avoided. Professional stereographers have a few tricks for fixing window violations with lighting or soft masks but it is best for beginners to simply obey this rule.

Fig. Rotational and Vertical Disparities in Source Footage

Fig. Disparities Corrected so all Straight Lines are Parallel

Unwelcome Disparities (Geometric, Lens and Temporal)

Disparity is a "dirty word" for stereographers. In fact the only "good" type of disparity in S3D is horizontal disparity between the left and right eye images. As mentioned before, this is known as parallax. Any other type of disparity in your image (vertical, rotational, zoom, keystone or temporal) will cause the viewers eyes to strain to accommodate. This can break the 3D effect and cause muscular pain in the viewer's eyes or even nausea. Every stereographer will strive to avoid these disparities on set by carefully calibrating the steroescopic rig and it will be tweaked ever further in post production through the use of 3D mastering software.

Ortho-stereo, Hyper-stereo & Hypo-stereo

I already mentioned that the average interocular of humans is considered to be about 65mm (2.5 inches.) When this same distance is used as the interaxial distance between two shooting cameras then the resulting stereoscopic effect is typically known as "Ortho-stereo." Many stereographers choose 2.5" as a stereo-base for this reason. If the interaxial distance used to shoot is smaller than 2.5 inches then you are shooting "Hypo-stereo." This technique is common for theatrically released films to accommodate the effects of the big screen. It is also used for macro stereoscopic photography.

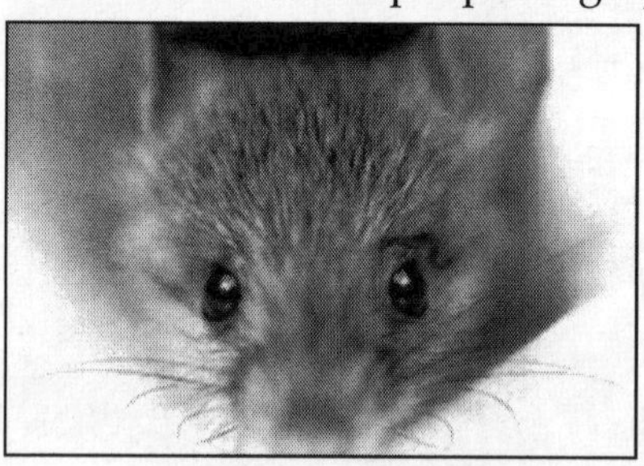

Fig. Hypo-stereo & Gigantism: Imagine how Objects Look from the P.O.V. of a Mouse.

Fig. Hyper-stereo & Dwarfism: Imagine how objects Look from the P.O.V. of an elephant.

Lastly, Hyper-stereo refers to interaxial distances greater than 2.5 inches. As I mentioned earlier the greater the interaxial separation, the greater the depth effect. An elephant can perceive much more depth than a human, and a human can perceive more depth than a mouse. However, using this same analogy, the mouse can get close and peer inside the petals of a flower with very good depth perception, and the human will just go "cross-eyed." Therefore decreasing the interaxial separation between two cameras to 1" or less will allow you to shoot amazing macro stereo-photos and separating the cameras to several feet apart will allow great depth on mountain ranges, city skylines and other vistas.

The trouble with using hyper-stereo is that scenes with gigantic objects in real-life may appear as small models. This phenomenon is known as dwarfism and we perceive it this way because the exaggerated separation between the taking lenses allows us to see around big objects much more that we do in the real world. Our brain interprets this as meaning the object must be small. The opposite happens with hypo-stereo, where normal sized objects appear gigantic. (Gigantism.)

If one attempts to shoot with two cameras configured in a side-by-side stereoscopic mount the smallest interaxial distance available will be the width of the camera. In most cases the width of the camera will be around 6 inches. This might seem like a big limiting factor, but other specialized equipment is available to achieve small interaxial distances with almost any sized camera. (More on that a in the "Selecting your Gear" segment.)

Viewing 3D: Passive Polarization, Active Shutter Glasses, Anaglyph & Autostereo

There are three basic types of glasses used for presenting stereoscopic 3D material. In most of the theatres in North America the common method is passive polarized glasses with either circular or linear polarizers. There are a few consumer and professional HD 3D monitors that use the same passive method. However, most of the consumer 3DTVs on the market use some form of active shutter glasses to flicker the left and right images on and off at 120Hz.

Autostereoscopic displays use lenticular lenses or parallel barrier technologies to present stereoscopic material without the use of glasses. Anaglyph glasses will work with almost any display but use color filters to separate the left and right images. The most common configurations are red/cyan, blue/amber, and green/magenta.

The Quick Math & Some Rules to Remember

Stereoscopic Parallax Budget (sometimes called Depth Budget) vs Depth Bracket

The Depth Bracket of your scene refers to the actual distance between your closest object in the frame and the furthest object. The Parallax Budget refers to your calculated maximum positive parallax and desired maximum negative parallax represented in percentage of screen width. For example if I determine through a simple calculation that my positive parallax should never exceed 0.7% of screen width and I have determined that my negative parallax should not exceed 2% of screen width, then my total Parallax Budget is 2.7%. The Depth Bracket must be able to be squeezed into the Parallax Budget. There are many algebraic formulas to determine the proper interaxial distance to achieve this.

Native Parallax for Final Display Size

The native parallax for a given screen size simply refers to what percentage of screen width will equal the human interocular. If you are using 2.5 inches as the baseline interocular and you know your presentation screen will be 30 feet wide (360 inches) then just divide 2.5 by 360. 2.5 ÷ 360 = 0.007 or 0.7% Therefore the Native Parallax of a 30 foot screen is 0.7%, so we should make sure to keep our maximum positive parallax under 0.7% of screen width if we plan to show our footage on a 30 foot wide screen. If we shoot for a 65″ 3DTV, then we can get away with over 3% positive parallax.

The 1/30th Rule

The 1/30 rule refers to a commonly accepted rule that has been used for decades by hobbyist stereographers around the world. It basically states that the interaxial separation should only be 1/30th of the distance from your camera to the closest subject.

In the case of ortho-stereoscopic shooting that would mean your cameras should only be 2.5″ apart and your closest subject should never be any closer than 75 inches (about 6 feet) away.

$$\text{Interaxial} \times 30 = \text{minimum object distance}$$

or

$$\text{Minimum object distance} \div 30 = \text{Interaxial}$$

Fig. If you are Using a Couple Standard 6″ wide Camcorders in a Side by Side rig as Close as they will Fit together then the Calculation would Look like: 6″ × 30 = 180 inches or 15 feet. That's right… 15 feet!

But does the 1/30 rule apply to all scenarios? No, the 1/30 rule certainly does not apply to all scenarios. In fact, in feature film production destined for the big screen we will typically use a ratio of 1/60, 1/100 or higher. The 1/30 rule works well if your final display screen size is less than 65 inches wide, your cameras were parallel to each other, and your shots were all taken outside with the background at infinity. When you are ready to take the next step to becoming a stereographer you will need to learn about parallax range and the various equations available to calculate maximum positive parallax (the parallax of the furthest object,) which will translate into a real-world distance when you eventually display your footage. Remember that illustration on the eyes pointing outward (diverging)? Well it isn't natural for humans to diverge and therefore the maximum positive parallax when displayed should not exceed the human interocular of 2.5 inches (65mm.) You can readjust the convergence point and bring the maximum positive parallax within the limits of the native display parallax (2.5 inches) but that will also increase your negative parallax.

Selecting Your Gear

Side by Side Rig vs Beam-Splitter Rig

Fig. Side by Side RIg

Fig. Beamsplitter RIg

Interaxial separation is an important factor when shooting S3D so therefore the width of your two cameras will determine the minimum interaxial separation in a side by side rig. Both of these interaxial distances are far too wide for any application other than hyper-stereo shots of landscapes, mountain ranges, helicopter shots, etc.

In order to shoot subjects in close range (within 15 or 20 feet) you will require a beamsplitter rig. Beam-splitters use a 50/50 or 60/40 mirror (similar to teleprompter glass) that allows one camera to shoot through the glass and then other to shoot the reflection. The interaxial can be brought down to as little as 0mm (2D) with beamsplitter rigs.

There are over 20 different beamsplitter rigs on the market ranging from $2500 USD to $500,000. However, many other types of disparity can be introduced when shooting through the glass (polarization effect, dust contamination, color cast, etc.)

Fig. Panasonic's H-FT012 for Micro4/3 Cameras

Fig. Loreo3D Attachment for DSLR Cameras

Special Stereoscopic Lenses

There are special stereoscopic lenses on the market designed for various digital SLR cameras. These lenses will work with a single camera but capture a left and right point of view in the same frame. The concept is intriguing but the lenses are very slow (F/11 – F/22), they use a smaller portion of the image sensor for each eye, they are usually made from plastic optics instead of glass and (in the case of the Loreo) the aspect ratio is vertically oriented.

Fig. Fujifilm's W1 S3D Camera

Purpose-built Stereoscopic Cameras

Stereoscopic film cameras have existed for decades. I personally own a Kodak Stereo camera from the early 50's that I've shot hundreds of 3D slides with and I love the simplicity. Recently manufacturers like Fujifilm, Panasonic, Sony and JVC have recognized the demand for digital versions of these cameras and released new products to market.

Some can record to separate left and right files or side-by-side format files for easy workflows in most non-linear editing systems (and easy compatibility with Stereo3D Toolbox) but many of the new systems record the two streams into a self-contained multi video codec (MVC) file that requires specific editing software (currently only Sony Vegas 10 on Windows) or a demuxing stage to separate the MVC into discreet left and right files (as with JVC's bundled Mac/PC software.)

Fig. JVC's GY-HMZ1U 3D Camcorder can Record Side by Side AVCHD (60i) or MVC (60i & 24p) and Ships with Mac/PC Demuxing Software

Fig. AJA's Gen10 Tri-Level Sync Generator and Hi5-3D Muxer

Genlock Capability

If you plan to shoot stereoscopic video with any action then it will be beneficial to use two cameras that can be genlocked together. Cameras that cannot be genlocked will have some degree of temporal disparity. However using the highest frame rate available (60p for example) will reduce the chance of detrimental temporal disparity. There are also some devices capable of synchronizing cameras that use LANC controllers.

Interlace vs. Progressive

Every frame of interlaced video inheritably will have some degree of temporal disparity between the fields. It is recommended to shoot with progressive formats whenever possible.

Lens & Focal Length Selection

Wider lenses will be easier to shoot with for the beginner and will also lend more "dimensionality" to your subjects. Telephoto lenses will compress your subjects flat so they appear as cardboard cutouts. Stay away from "fisheye" lenses because the distortion will cause many geometrical disparities.

OK, so you've learned your terminology and selected your gear. Now what? It's time to get out there and shoot. We haven't discussed the various calculations or the rules of S3D but I encourage you to shoot now so you can learn from your mistakes.

Turn off Image Stabilization

If you are using video cameras with image stabilization you must turn the feature off or the camera's optical axis will move independent of each other in unpredictable ways. As you can imagine this will make it impossible to tune out disparities.

Manually Set White Balance

Use a white card, chart or 18% gray card to set the manual white balance of both cameras. On beamsplitter rigs it is not advisable to use preset white balance settings because the mirror glass introduces its own tint to the image on each camera. Set the WB switch to either A or B and press and hold the AWB button to execute the white balance sequence.

Gain

It is best to shoot on 0dB gain when possible. The noise and grain patterns at high gain levels will be unique on each camera for each frame and therefore will be a visual disparity.

Use Identical Settings on both Cameras

It is very important to use the same type of camera, same type of lens and exactly the same camera settings (white balance, shutter speed, aperture, frame rate, resolution, zoom, codec, etc.) on both cameras. Any differences will cause a disparity. It is also a good idea to use manual focus and set it to the hyperfocal distance or a suitable distance with a deep depth of field.

Proper Configuration for CMOS Shutters

Fig. CMOS Sensor Cameras in Proper Configuration

The CMOS sensors in cameras like the Sony F3, Red, Canon XF105 or XF305 use a rolling shutter that requires a particular mounting configuration in a beamsplitter rig.

The tops of the frames must match so there is no rolling shutter disparity between the sensors. If the mirror in your rig faces the ground and the upright camera mounts underneath then the camera can be mounted on the plate normally. If your mirror faces up and the upright camera points down then the camera must be mounted upside down so that the top-bottom orientation of the sensors match.

Use a Clapper or Synchronize Timecode

If your cameras are capable of genlock and TC slave then by all means use those features to maintain synchronization. If you are using consumer level cameras it will be up to you to synchronize the shots in post. In either case you should use a slate with a clapper to identify the shot/takes and easily synch them. If your cameras have an IR remote start/stop it is handy to use one remote to roll & cut both cameras simultaneously. If you are shooting stills with DSLRs there are ways to connect the cameras with an electronic cable release for synchronized shutters.

Slow Down your Pans

However fast you are used to panning in 2D, cut the speed in half for 3D. If you are shooting in interlace then cut the speed in half again. Better yet, avoid pans altogether unless your cameras are genlocked. Whip pans should be OK with genlocked cameras.

Label your Media "Left" and "Right"

This might seem like a simple rule to remember but the truth is that most instances of inverted 3D is a result of a mislabeled tape or clip. Good logging and management of clips is essential with stereoscopic post production.

To Converge or Not Converge... That is the Question

One of the most debated topics among stereographers is whether to "toe-in" the cameras to converge on your subject or simply mount the cameras perfectly parallel and set convergence in post-production. Converging while shooting requires more time during production but one would hope less time in production.

However "toeing-in" can also create keystoning issues that need to be repaired later. My personal mantra is to always shoot perfectly parallel and I recommend the same for the budding stereographer.

Post

So you've shot your footage and now you want to edit and watch it. If you work with After Effects, Motion or Final Cut Pro on the Mac please watch some of the tutorials on this website to learn more about how Stereo3D Toolbox can help you master your S3D content.

Fixing Disparity and Setting Convergence

Most stereoscopic post software has sliders to adjust vertical, rotational, zoom, color & keystone disparities. Fixing these disparities requires skill and practice but my recommendation is to start with rotation and make sure any straight lines are parallel to each other and then adjust zoom to make sure objects are the same apparent size. Next adjust the vertical disparity control make sure all objects next to each other. Finally adjust the horizontal convergence to perfectly align the object you wanted to be on the stereo window.

Native Pixel Parallax

There is one last thing you should check after aligning each shot. You must make sure that your background doesn't exceed the Native Pixel Parallax of your display screen or your audience's eyes will diverge (which is bad.) The idea here is that the maximum positive parallax (the parallax of your deepest object/background) does not exceed the human interocular distance when presented. You can determine the Native Pixel Parallax (a.k.a. NPP) by dividing 2.5 inches by the display screen's width and then multiply the result by the amount of horizontal pixels.

I present my S3D material on JVC's 46" 3DTV. It is 42 inches wide and 1920 pixels wide so the calculation is 2.5/42 × 1920 = 114 pixels. This means that the parallax of the background should not exceed 114 pixels.

In Stereo3D Toolbox you can enter your screen width and the filter will automatically calculate NPP and display a grid. If the parallax in your background does exceed this limit then adjust your convergence to move the depth range back away from the viewer.

Share your S3D Masterpiece on YouTube with the yt3d Tag

Now that you have finished editing and mastering your S3D movie it is time to share it with the world. YouTube added the capability to dynamically present S3D content in any anaglyph format. All you have to do is export your movie file as "side by side squeezed" and encode it as H264 with Compressor. I recommend using 1280x720p for S3D content on Youtube but not 1080p. The workload of rendering the anaglyph result is handled by the viewer's computer so 1080p will decrease the frame rate on most laptops.

Fig. Output as Side by Side Squeeze

Upload your movie file to YouTube and then add the tag "yt3d:enable=true" to enable YouTube 3D mode. If your footage is 16×9 aspect ratio also add the tag "yt3d:aspect=16:9". YouTube 3D expected crossview formatted side by side so if you exported as side by side parallel instead of crossview you will need to add the tag "yt3d:swap=true" to ensure the left and right eyes are presented correctly. Remember stereoscopic 3D cinematography is a craft that can takes years to master and is a craft where even the 'experts' are still learning new techniques. As the popularity of S3D continues to rise there will by many demands on inexperienced videographers to provide stereoscopic services. It is important to remember that 2D can look bad, but 3D can feel bad. The last thing any producer wants is to physically hurt the audience. Therefore, extensive practice and testing is advised before producing content to be viewed by anyone other than the stereographer. Trial and error is the best way to learn this particular craft.

6

Public Sphere and Modern Television

Television in its many senses – as a medium, an institution, an industry – has gone through fundamental changes in the past decade. These, in turn, have evoked numerous predictions and scenarios, from 'dumbing down' of television culture to the eventual 'death of television', inspired by the technology-oriented utopias of the Internet taking over as the key medium, and television merging with other forms of media. Still, TV remains as the most influential mass medium in Europe, alone if measured simply by the time people spend with it. In fact, there are great expectations for television as the key medium enhancing democracy and civic participation, expressed in the European Union policy statements as well as in national memoranda and reports on the topic.

Yet, for many media scholars as well as political scientists, Jürgen Habermas' theory of the public sphere may at the outset appear outdated when considering today's television landscapes in Europe or elsewhere. In his famous study The Structural Transformation of the Public Sphere (1989/1962), Habermas builds his model upon the development of the bourgeois society of the 17th and 18th century Europe. Based on that context, he suggests that a space was then formed and realised between the economy and polity where people could be informed and discuss, so as to form decisions and act upon them. The instruments of this sphere were newspapers, books, salons and debating societies that allowed an arena relatively separate of the Church and the State, characterised by openness to all citizens, in the sense of the 18th century. As follows, he sees such a communicative forum as a model for a public sphere that facilitates rational-critical debate amongst citizens, that accordingly ensures political will formation, and that is therefore crucial for democracy.

In contrast, the present context, with technological advances such as digitalisation, coupled with debates on globalisation as well as on cultural, economic and political neoliberal tendencies might point towards other approaches in understanding television's role in the society. To be sure, much of postmodern theorisation of popular culture in the 1990s would suggest alternative outlooks. Also, Habermas himself considered the public sphere already at the time of the publishing of his study to be 'refeudalised'. He viewed the mass media's role as a public sphere dubious, as they give in to commercial imperatives and address their viewers as consumers. His scepticism is illustrated by the following quotes:

"Mass culture has earned its rather dubious name precisely by achieving increased sales by adapting to the need for relaxation and entertainment on the part of consumer strata with relatively little education, rather than through the guidance of an enlarged public toward the appreciation of a culture undamaged in its substance."

"The world fashioned by the mass media is a public sphere in appearance only. By the same token the integrity of the private sphere which they promise to their consumers is also an illusion."

Yet, as the above citations illustrate, his argumentation seems to share features with the public debate that surrounds at least European television today. These arguments address the concern for contents geared towards entertainment for commercial reasons, for the positioning of viewers as indulging consumers rather than rational-critical citizens, for the diminishing political participation and, as the end result of these tendencies, for the crisis of democracy.

The resonance and relevance of the Habermasian idea of the public sphere can also be said to manifest in the vast amount theorising and application it has inspired. Alone the amount of revising and criticism emphasises the importance of this thought. Different points of dissatisfaction include the claim that Habermas idealizes the bourgeois public sphere and its thrive for public enlightenment while neglecting other, 'plebeian' public spheres. It has also been said that he exaggerates the idea of 'refeudalisation' and the manipulative powers of culture industries in the mass media era. Yet another critique is that Habermas emphasises rational consensus rather than the continuing need for compromise of differing interests, and for a pluralist public sphere. Thus, it has been claimed, he dismisses communication that is not geared toward consensus and accordingly, creates too great of a boundary between information and entertainment.

Despite of the doubts presented about the public sphere idea, Habermas' work has influenced many disciplines and debates ranging from the democratic theory of political science, the self-reflection of cultural critics, the modernism/postmodernism debate, to empirical studies in sociology and communications. Specifically, as Peter Dahlgren notes, the role of television has explicitly or implicitly been framed in the public sphere thematic also in countless recent analyses, whether they concern new genres such as those that fall under the term of reality television, whether they address globalisation of national television culture, or whether they seek to tackle the evolution of broadcasting journalism. The public sphere has proven useful both as an analytical concept but also as normative ideal – and, as Dahlgren (op cit.)

points out, it has in the course of years left the strict Frankfurt School theory realm and is now often used more generically to refer to democratic goals and responsibilities of the media and the civic life.

It is in the spirit of the latter definition that Peter Dahlgren (2004) writes about the role of theory as 'intellectual scaffolding':

"[Theory] serves to orient us, to pull together sets of facts and assumptions, and offers normative dispositions. It helps to provide significance to what we observe, and to suggest the various types of action or intervention. In this sense, there may not always be demarcated distinctions between formalized theory and the more general (and less systematic) thought modes we use." Still, in this paper I do not advocate on the notion of the public sphere solely as a vague set of expectations for the media in democratic societies, but wish to sharpen the idea with the help of existing criticism as well as specify it regarding television. The thematic of the public sphere, I argue, benefits from the various further theorizations and coupled with them, may further understanding of the phenomenon of television of the past decades as well as today. The main features of the Habermasian public sphere that call for revision today, are therefore discussed below.

NORMATIVE SPHERE REVISITED

First, in the most fundamental level, most critics seem to agree that Habermas' model is too normative as it celebrates the 18th century liberal world. In the realm of theorisation on democracy, it could be interpreted that the existence of a public sphere in the 20th century social-welfare mass democracy is virtually impossible. Habermas, then, sees little potential with today's mass media that are not only distorted by commercial imperatives to entertain, but that also reduce to a minimum the 'distance that required the privacy of appropriation as much as it made possible the publicity of a rational-critical exchange'.

Regardless, there is no question about the fundamental role that is given to the notion of the public sphere when defining the media's role in the society. In this context, radio and television – and beginning of the new millennium also new media such as the Internet – cannot be dismissed. The electronic media have become undeniably and tightly connected with what is considered political, both in traditional institutional sense as well as an integral part what cultural studies oriented scholar call the politics of the everyday and politics of popular culture. Accordingly, it has been argued that the emphasis on the interpersonal communication as the normative basis for the public sphere must be re-evaluated, and meanings of time and space for public communication have to be addressed. The discussion is given another twist by the cultural studies media theorists who address the media in the postmodern context. Thinkers of this tradition recognise and dismiss Habermas' normativity and see his work too deeply rooted in the modern Enlightment philosophies.

What is Public, What is Private

Linked to the above critique is another key notion embedded in the idea of the public sphere, namely the distinction or the opposition between the public and the private. This have been extensively theorised and debated within political and social sciences. Alone the definitions of the public and the private are tricky: For example, John B. Thompson distinguishes two different meanings for private and public domains which both bear relevance in discussions about the media. The first meaning refers to the private domain as consisting of organisations functioning in the market economy and the personal life and relations; and the public domain including state-owned economic and other organisations. In between exist, in his view, intermediate organisations such as political parties. On the other hand, Thompson sees another meaning in the concepts of the public and the private that deal their meanings as attributes (public, private): These are related to visibility and invisibility, publicness and privacy.

The latter meaning has been widely discussed in relation to the media, both by political scientists and cultural studies scholars in relation to the media, as it seems that the conventions regarding these attributes are rapidly and continuously changing. Often used empirical examples are politicians' celebrity status and their private affairs making headlines in 'quality' as well as popular media; and the mirror phenomenon of 'ordinary people's' lives becoming public in talk shows and reality programmes. In most Western democracies, worried discussions have been carried out on the redefinitions of the public and the private, the fear being that the increasing interest in the private and the personal overshadows issues of more general social and political importance.

One of the frequently cited works from the perspective of political communication has even announced the crisis of public communication. Other vain of thinking is that the new conventions of the public and the private challenges the idea of the rationally functioning citizen and bring about new definitions of the political and possibilities of bringing the formerly private into the public realms. Some authors have even claimed that the kind of politics that truly counts is formulated and acted upon in the private sphere and everyday life; and that such politics are fuelled best if not solely by popular culture which evokes resistance against the official hegemony or 'power-blocks'. Whether either of these views is accurate is one question, but they highlight the notion that the Habermasian public sphere need further consideration also in the definition of what is meant and desired for by the term public.

Exclusive Public Sphere

Related to the need for updating the notions of the public and the private is the more specific assessment, made by many scholars in the 1990s, that

Habermas remains insensitive to the dynamics of gender or other struggles of subordinate social groups. This vain of thinking finds ground both in critical theory and in cultural studies. Notably, feminist political scientists such as Nancy Fraser and media scholars such as Lisa McLaughlin and Annabelle Sreberny-Mohammadi and Liesbet van Zoonen have emphasised the patriarchal nature of the inspiration of the model, the bourgeois public sphere, which excluded women from its scope of citizenry. This is certainly true for many other social groups and movements, and this is why Habermas has been accused of élitisim. The feminist critique, however, relates to the discussion on tabloidization debate of e.g., personalization of traditional politics and intimization of journalism as it recognises the rigid separation of public and private spheres, embedded in the Habermas' ideal, that emphasises the traditionally gendered division of those spheres.

This critique, reflecting a strand of academic work that flourished in the last decade, can be elaborated by paying attention to the emergence of new kind of civic activism, for instance in the form of loose networks and single-issue movements. The idea of a public sphere, provided by one medium (or even by several media in unison), fostering one kind of civic culture, is clearly outdated alone because of the factual reality of ever fragmenting media audiences, of different kinds of contents within one medium and different needs to follow different kinds of contents, and of the different uses of old and new media. So while for example television is not dead, it certainly cannot be said to cater for the public sphere, for everyone simultaneously.

PUBLIC SPHERE AND PUBLIC SERVICE

There is a clear connection between the public sphere theorization and public service broadcasting. On one hand, this is a paradox, as Habermas' model seems to advocate a neutral sphere that defies both the corporate and state control. From a slightly different angle, one of the model's strengths can be found precisely in that it escapes the trap of the traditional opposition of free markets versus state-governed media. On the other hand, the normative nature of the model seems to translate well into media policy practices and debates and to adjust to the mission of public service broadcasting.

Undeniably, Habermas has been used to discuss the public service broadcasting mission and for its defence, against deregulatory and commercialising tendencies in practice. Concerning European media policies, a persistent (if latent) sense of the need to construct a public sphere has guided many public service media organisations. Deregulation has in the past years changed rhetorics of media policies in the EU level towards consumer-oriented discussion, and a similar trend can be seen in Finland, for instance regarding digitalization: The ideals of the free market and consumer's choice are present in the public rhetoric of the Ministry of Traffic and Communications and similarly, the directors of Finland's public service broadcasting company YLE tend in the new millennium refer to audiences as consumers. Still, the ideals of the public sphere are present, for instance, in many aspects of the law on YLE, such as the ideals of free access to all citizens, and the like. This discussion is naturally linked to the claims of crisis of public communication - involving aggressive public relations and tyranny of the media as well as decline of journalism - that threatens many liberal democracies. Similar vain of thought has also been expressed regarding globalizing media markets: Rich media seems to result in poor democracy.

Controversy and the Political

When controversial issues regarding media contents such as 'trash television' or 'entertainization of TV culture' are addressed in the academic and/or public discussion, they are often presented together with the above mentioned trends. For this reason I argue that in addressing popular media culture, the public sphere paradigm and the post modern cultural theory have meetings points in searching for articulations of the political. Or, as Kevin Glynn explains his motivation to study tabloid culture:

"Wherever the cultural tastes and practices of some people disgust and offend others, there can be little doubt that we are in the presence of the political... The production of disgust, offence, and popular pleasures is of primary significance for cultural theory because it is central to the general process whereby the meanings we make ourselves and of the social world are organised and reorganised. This process has crucial implications for social change, not least because it shapes the production, diffusion and urgency of the recognition of a need for change."

In other words, if politics is about different (social) interests for which solutions are sought, the public sphere paradigm and the (some forms of) cultural theory seem to agree in the basic assumptions and goals. In this light, it could be argued that the public sphere constructed by debate and discussion is not necessarily oppositional to discursively constructed realities, a stand point advocated in post-modern theorisation. As Fraser argues, Habermas' public sphere can be seen as:

" [A] theatre in modern societies in which political participation is enacted through the medium of talk ... a space in which citizens deliberate their common affairs, and hence an institutionalised arena of discursive interaction".

Thus, it is symptomatic that the public sphere framework has been recognized and even utilised alongside the cultural studies oriented readings on tabloid television and tabloid journalism in general.

In sum, the thematic of the public sphere as theoretical inspiration or scaffolding for understanding today's television is relevant for several reasons. First, the academic and public debate on changes in the television landscape can traced to the public sphere ideal. I argue that the public sphere model is an interesting starting point especially when examining European television systems that are based on a strong public service and hence public sphere ethos, but relevant also in any case with nation-bound television. Lastly, Habermas' model provides not only a theory but a normative stand. Sonia Livingstone and Peter Lunt (1994) who have used the public sphere paradigm when examining a form of tabloid television, that of talk shows, aptly articulate the promise that just may have caused Habermas' appeal to a variety of theoretical and empirical applications in the past decades:

"Habermas' position reflects the ambivalence felt by many towards the mass media – that there is a great power, but can it be harnessed for the public good? We suggest that pessimistic answers tend to underestimate the complex and contradictory or fragmented nature of the contemporary mass media which opens the way for some escape from institutional control, while more optimistic positions often set too high ideals for the public sphere. Those alternative formulations of the public sphere which recognize and build on the complex and fragmentary nature of the media suggest more positively that the media could facilitate and legitimize the public negotiation – through compromise rather than consensus – of meanings among oppositional and marginalized groups."

THE PUBLIC SPHERE

Jürgen Habermas' seminal The Structural Transformation of the Public Sphere provides a basis for discussions of the public sphere/s in contemporary societies. In the following I attempt to summarize Habermas' thesis and place it within the framework of recent debates on the public sphere/s and media.

In his historical narrative informed by the Frankfurt School tradition Habermas traces the development of the bourgeois public sphere and its consequent transformation. According to him, the bourgeois public sphere reached its peak in the early to mid-19th century. Habermas argues that the public sphere emerged as a space in which private individuals came together as a public to use their own reason to discuss the power of the state. The bourgeois public sphere thus came into existence as a result of struggle against despotic states. The development of competitive market capitalism led to the creation of institutions within civil society that occupied a space distinct from both the economy and the state. These institutions included newspapers, debating societies, salons and coffee houses. Although Habermas understands the bourgeois public sphere of the early 19th century as an ideal model, a peak of rational discussion, he identifies problems with its universalism. To be part of the public at the time meant to belong to the property-owning class and to be literate. Habermas also points out that the emergence of the bourgeois public sphere went together with an institutionalization of privateness, with the constitution of a clear distinction between public and private. With the height of laissez-faire capitalism the public sphere underwent significant, and according to Habermas detrimental, changes. The decline of the public sphere is connected with rapid social developments, industrialization, urbanization, growth of literacy and popular press and other related factors. These changes, to use Dahlgren's summary, resulted in "a blurring of the distinction between public and private in political and economic affairs, a rationalization and shrinking of the private intimate sphere (family life) and the gradual shift from an (albeit limited) public of political and cultural debaters to a public of mass consumers." The mutual penetration of state and society (which Habermas terms refeudalization) dissolved a private sphere, also the basis for a relatively homogeneous public composed of private citizens engaged in a rational-critical debate was threatened and competing organized private interests invaded the public sphere.

The rational-critical debate that characterized the bourgeois public sphere at its peak was replaced by consumption (or such a debate is entirely shaped by the media) and for Habermas it continues to exist as a public sphere in appearance only. Further changes in the public sphere followed with the emergence of the welfare state. Habermas argues that public opinion is no longer the result of rational debate but the outcome of media engineering. The public sphere has become a platform for advertising and the press has become mere trade. In order to restore the function of the public sphere, Habermas proposes a discourse-centred theory of democracy, according to which a majority decision must be a "rationally motivated but fallible result of a discussion concerning the judicious resolution of a problem, a discussion that has come temporarily to a close because coming to a decision could no longer be postponed". Thus a political public sphere would be characterized by at least two crosscutting processes: communicative generation of legitimate power and a deployment of media power to procure mass loyalty, consumer demand and compliance with systematic imperatives.

This brief summary of Habermas' arguments does not by any means aim at providing a full account of his thesis. Rather, my aim is to draw attention to the role that Habermas attributed to mass media in the disintegration of the public sphere and to discuss whether he understands mass media as playing any other than a negative/destructive role in relation to the public sphere. In the summary I have already alluded to Habermas' views on the press, he identifies a particular problem which is a consequence of the conflation of journalism and literature and results in conjuring a peculiar reality, even a conflation of different levels of reality: "instead of doing justice to reality, [journalism] has a tendency to present a substitute more palatable for consumption and more likely to give rise to an impersonal indulgence in stimulating relaxation than to a public use of reason". However bleak this picture may seem, it was to get worse with the emergence of radio and television.

With the arrival of new media [radio and television] the form of communication as such has changed; they have had an impact, therefore, more penetrating (in the strict sense of the word) than was ever possible for the press. Under the pressure of the "Don't talk back!" the conduct of the public assumes a different form. In comparison with printed communications the programs sent by the new media curtail the reactions of their recipients in a peculiar way. They draw the eyes and ears of the public under their spell but at the same time, by taking away its distance, place it under "tutelage," which is to say they deprive it of the opportunity to say something and to disagree. The critical discussion of the reading public tends to give way to "exchanges about tastes and preferences" between consumers – even the talk about what is consumed, "the examination of tastes," becomes part of consumption itself.

In more concrete terms Habermas identifies the degree of economic concentration and technological-organizational co-ordination in media as a threat to the critical functions of publicist institutions. In respect of this problem he points out that due to the high degree of concentration, governments often opted for putting media under public control rather than private ownership. "Thus the original basis of the publicist institutions, at least in their most advanced sectors, became practically reversed. According to the liberal model of the public sphere, the institutions of the public engaged in rational-critical debate were protected from interference by public authority by virtue of their being in the hands of private people".

Habermas' concept of the bourgeois public sphere and its transformation has been criticized mainly on three grounds. Firstly, although Habermas admits the exclusionary nature of the bourgeois public sphere in terms of class, he omits the question of gender altogether. Secondly, he remains silent on alternative public spheres. In this respect the work of Oskar Negt and Alexander Kluge provides an interesting comparison, with their notion of a proletarian counter-public sphere. However, Polan convincingly argues that Negt and Kluge tend to idealize this counter-public sphere similarly to Habermas' idealization of the bourgeois public sphere. And finally, as Peter Dahlgren points out, Habermas' Structural Transformation of the Public Sphere is characterized by an "absence of reference to the complexities and contradictions of meaning production" as well as "to the concrete social settings and cultural resources at work".

Habermas himself acknowledges that some of his arguments need revision and in particular that his "diagnosis of a unilinear development from a politically active public to one withdrawn into a bad privacy, from a 'culture-debating to a culture-consuming public' is too simplistic". Yet, he maintains that the bourgeois public sphere as it existed in the early 19th century provides an ideal model, arguing that a single political public sphere that is built on the principles of communicative action can serve contemporary societies.

Nicholas Garnham stresses two shortcomings in relation to Habermas' understanding of the public sphere. First, the relation of the concept to the historical reality of class division and class politics, he, however, points out that despite the validity of this criticism we do not need to give up on the universalistic possibility of a general public sphere. Secondly, argues Garnham, Habermas conceives of public sphere on an individualistic basis, "an assumption, which seems to me wholly unrealistic, that all participants possess complete information and engage in all debates," consequently the concept cannot at its present form deal with the political problem of mediation and equally there is no space left for the social role and power of expertise and expert knowledge nor for the role and social interests associated with knowledge broking. "Thus it becomes difficult to handle the problem of the role of those who in fact manage the conduct of the information-gathering and debate which is the Public Sphere's raison d'étre, namely, in particular, journalists and politicians themselves. It is a further result of this weakness that the theory has no place for what I regard as an essential and central organizing institution within the Public Sphere, the political party".

Despite these criticisms Garnham maintains that three key components of Habermas' thesis are essential in an understanding of the public sphere/s in contemporary societies. Firstly, Habermas focuses on "the indissoluble link between the institutions and practices of mass communication and the institutions and practices of democratic societies". Secondly, he stresses the "necessary material resource base for any public sphere" and finally Garnham praises Habermas for his avoidance of the simple dichotomy of free market versus state control in his distinction of the public sphere from both state and market that enables him to discuss the question of threats to democracy from both of them. Garnham believes that these arguments have particular relevance in contemporary societies as the emergence of global markets as centres of private economic power undermines the nation state and new public spheres and political institutions are needed for the control of the global polity and economy. Moreover, he goes on to argue, public broadcasting institutions are themselves undergoing changes with the development of new media and the progressive destruction of public service media.

Nancy Fraser's critique of Habermas' argument relates closely to the issues raised by Garnham. She argues that revisionist historiography questions four assumptions central to Habermas', "bourgeois masculinist", understanding of the public sphere. Firstly, it challenges the assumption that "societal equality is not a necessary condition for political democracy". Secondly, it questions that a multiplicity of public spheres is a step away from democracy. Thirdly, it undermines the assumption that discourse in public spheres is devoted solely to public good, leaving private interests and issues undesirable. And finally, the assumption that a functioning democratic public sphere requires a sharp separation between civil society and state is not necessarily a valid one.

Regarding the notion that rational debate in the public sphere should promote public good, Fraser suggests that there should be no restrictions on topics debated in the public sphere as there is no guarantee that the outcome of a deliberative action will be the discovery of common good.

One of the central questions that has been explored by critics of Habermas' argument is whether there is or has ever been a single public and consequently a single public sphere. Bruce Robbins convincingly argues that historical narratives (Habermas' among them) in which an idea of a lost public is evoked miss a crucial point.

The appearance of the public in these historical narratives is something of a conjuring trick. For whom was the city once more public than now? Was it ever open to the scrutiny and participation, let alone under the control, of the majority? Was there ever a time when intellectuals were really authorized to speak to the people as a whole about the interests of the people as a whole? If so, where were the workers, the women, the lesbians, the gay men, the African Americans?

More insight can be gained if "one moves away from the universalizing ideal of a single public and attends instead to the actual multiplicity of distinct and overlapping public discourses, public spheres, and scenes of evaluation that already exist, but that the usual idealizations have screened from view".

Fraser argues that a single public sphere does not provide an arena for subordinated groups in which they can discuss their needs, objectives and strategies, "they would have no venues in which to undertake communicative processes that were not, as it were, under the supervision of dominant groups". The existence of multiple public spheres would enhance participation in the sense of "being able to speak in one's own voice, and thereby simultaneously to construct and express one's cultural identity through idiom and style". Moreover, public spheres are not spaces of zero-degree culture, they "consist in culturally specific institutions, including, for example, various journals and various geographies of urban space. These institutions may be understood as culturally specific rhetorical lenses that filter and alter the utterances they frame; they can accommodate some expressive models and not others". Privileging a single seemingly all-inclusive public sphere thus results not only in a reduction of these diverse modes of expression but indeed in a privileging of a dominant mode. In her conclusion Fraser states that what we ultimately need is "a critical political sociology of a form of public life in which multiple but unequal publics participate".

John Keane argues that there are empirical reasons alone why the concept of 'public spheres' should be brought to bear on phenomena as disparate as computer networking, citizens' initiatives, newspaper circulation, satellite broadcasting, and children playing video games.... Public spheres can and do develop within various realms of civil society and state institutions, including within the supposed enemy territory of consumer markets and within the world of power that lies beyond the reach of nation-states, the Hobbesian world conventionally dominated by shadowy agreements, suited diplomacy, business transactions, and war and rumours of war. While I agree with Keane's suggestion that multiple public spheres actually exist in contemporary societies, I would argue that it is not enough to explore existing public spheres in a redemptive fashion and categorize them (Keane distinguishes among micro, meso and macro public spheres) rather one needs to theorize and explore the power relations and conflicts within them and among them as well as how these are translated into political action. I agree with Nicholas Garnham who argues that by defining public spheres Keane abandons "the problem the term was designed to address – namely how can the promise of democracy be realized".

It is clear from the above that media are in the centre of discussions about the public sphere/s, yet there are also larger questions at stake such as allocation of resources, equality and justice. As Todd Gitlin rightly points out "if it be argued that a single public sphere is unnecessary as long as segments constitute their own deliberative assemblies, such an argument presumes a rough equivalence of resources for the purposes of assuring overall justice".

PUBLIC SPHERE AND PUBLIC SERVICE MEDIA

John Keane suggests that in contemporary societies "the public sphere ideal is linked to the institution of public service broadcasting, which is seen to have an elective affinity with public life and to be the best guarantee of its survival in the era of state-organized, consumer capitalism". Carpignano et al. argue that in the debates about mass media, politics and the public sphere "there is a common ground, a mutual acceptance of basic premises, shared by participating politicians, conservative ideologues, and leftist cultural critics. Its unquestionable truism is that the mass media today are the public sphere and that this is the reason for the degradation of public life if not its disappearance". Charles Husband argues that mass media are "a core element in civil society and a fundamental prerequisite for the promotion of civic trust in complex multi-ethnic societies".

In classical liberal theory media are understood as crucial in providing "the space between government and society in which private individuals exercise formal and informal control over the state: formal control through the election of governments and informal through the pressure of public opinion". The media play a central role in this process as they distribute information as well as facilitate the formation of public opinion and provide an independent forum for debate. "The media are thus the principal institutions of the public sphere or, in the rhetoric of nineteenth-century liberalism, 'the fourth estate of the realm'". It thus seems to be a widely accepted fact that mass media indeed are the public sphere.

The focus of this dissertation is broadcasting which was in Europe originally conceived as a service to the public, the spectrum was understood as a "public natural resource". According to Hall (1993) the public service idea clearly has its basis in the claim that there is "such a thing as 'the public interest' – a social interest – at stake in broadcasting", he goes on to identify some of the roles of broadcasting in modern societies (source of knowledge, creator of a discursive space, a key pass between "the governed" and "the governors") to argue that "access to broadcasting has thus become a condition, a sine qua non, of modern citizenship" (ibid: 25, original emphasis). From the very beginning thus an affinity was identified between media and the public, in order to clarify it I briefly discuss the notion of public interest in relation to communication.

According to McQuail the origin of the term can be found in economic regulation, Mitnick derives it from medieval social theory which led to ideas of economic justice supportive of collective control over market forces. In modern societies we find examples of sources that are regulated in the public interest, these include basic infrastructure as well as basic telecommunication services. Telecommunication services are part of businesses that are considered to be "affected with a public interest" due to the essential nature of the service, its tendency to monopoly and the requirement of universal accessibility. McQuail, however, notes that difficulties with applying the term public interest to an area like communication have their source in misunderstandings according to which features of mass communication are essential and whether interferences with free market mechanisms are justified in order to secure these.

The problematic of identifying public interest in communication is demonstrated by difficulties with finding its place in typologies of public interest. McQuail draws on Held's variants of public interest theory which distinguishes three of them. The first, preponderance theory, understands public interest in a majoritarian way, i.e. it aims at maximizing the number of individual preferences. The second theory, common interest theory, refers to cases "where the interests in question are ones which all members are presumed to have in common, with little scope for dispute over preferences". The third theory Held identifies is the unitary one, which is "in effect, the assertion of some absolute normative principle, usually deriving from some larger social theory or ideology". According to McQuail, it is solely the common interest theory type which can be used for identifying public interest in communication, the other two types are insensitive or irrelevant to some key issues (including popular wants).

It is probably not surprising that apart from difficulties with pinpointing public interest in broadcasting there are also various notions about how best to ensure that broadcasting serves the public. The debate over this question has been dominated by two opposing views. On the one hand we find advocates of privately-run commercial media systems who argue that this is the way of guaranteeing independence and the market will serve the interests of all consumers – it will provide them with what they want. However, as for example John Keane argues in The Media and Democracy, unrestricted competition does not necessarily ensure freedom of entry into the market place. Moreover, free market competition restrains the choices of some and advertising, which plays a central role in market competition, reduces the supply of minority interest programmes and tends to shut off non-commercial opinions and non-market forms of life.

In contrast, we find proponents of the view that public service media can best serve the interests of consumers as well as citizens. James Curran draws upon the work of neo-Keynesians critical of the free market in broadcasting. This critique states that public service media serve "public good" without incurring additional costs, that all advertising-funded markets are imperfect as they are biased against quality (favour high ratings as opposed to highly rated programmes) and favour majority programmes which renders them insensitive to intensities of demand. Moreover, the emergence of new technologies actually increases the need for public control, as one source of monopoly (spectrum scarcity) has been replaced by the monopoly of economies of scale and scope. Curran very rightly points out that an alternative approach to the question of public service broadcasting takes into account the fact that people are not only consumers but also citizens within a democratic system with a right to be adequately informed about matters relating to public interest. A right, Curran argues, that is best guaranteed by public service broadcasting because "it gives due attention to public affairs, and is less dominated by drama and entertainment than market-based broadcasting generally is".

Thus far my argument has involved theoretical issues concerning public service broadcasting and its proposed roles. Although in the following section I outline three types of public broadcasting systems currently existing in Western democracies it has to be noted that there are certain features that are shared by public service broadcasting in general. According to Siune and Hultén the "old order" of public broadcasting in Europe of the early 1990s shared the following defining elements: some form of accountability to the public, some element of public finance, regulation of content, universal service (in the territorial sense) and regulated entrance (i.e. limits to the number of competing channels).

In more abstract terms Garnham identifies the strengths of public service broadcasting in the following: "(a) [it] presupposes and then tries to develop in its practice a set of social relations which are distinctly political rather than economic, and (b) at the same time attempts to insulate itself from control by the state as opposed to, and this is often forgotten, political control". The now defunct Broadcasting Research Unit (from the UK) identified the principles of public service broadcasting as the following: universal accessibility (geographic); universal appeal (general tastes and interests); particular attention to minorities; contribution to a sense of national identity and community; distance from vested interests; direct funding and universality of payment; competition in good programming rather than for numbers; and guidelines that liberate rather than restrict programme makers. James Curran distinguishes three types of systems of public service media:

Social Devolutionary

We find this type for example in Italy and the Netherlands, according to McQuail it can be characterized by external diversity, i.e. various channels or time blocks are allocated to various interest groups within the society. To make the example clearer, I provide a brief description of the Dutch public broadcasting system which reflects the Dutch social system that can be summarized in one word: pillarization. "Dutch society between the beginning of the twentieth century and the mid-1960s (and notably the first 20 years after the Second World War) was a principal example of 'segmented pluralism', with social movements, educational and communication systems, voluntary associations and political parties organized vertically (and often cross-cutting through social strata) along the lines of religious and ideological cleavages". The Dutch public broadcasting system works on the principle of allocating access to associations with different outlooks and priorities. According to the law "a broadcasting association should aim, as laid down in its statutes, to represent some clearly stated societal, cultural, religious or philosophical stream and to direct itself in its programming to the satisfaction of some actively present social, cultural, religious or philosophical needs".

McQuail argues that the idea of diversity as expressed in the Dutch broadcasting system mainly relates "to an 'external' and exclusive diversity in which different 'voices' and outlooks have their own separate channels, rather than to the more commonly encountered 'internal' diversity, according to which all tastes are catered for by channels serving large, heterogeneous audiences". In practice, the "allocation of broadcasting time was based on the number of members and/or subscribers to the broadcasting magazines produced by the different organizations". Dutch public service broadcasting is financed by licence fee, advertising and membership dues and magazine subscription. Commercial broadcasting was legalized in 1990.

Liberal Corporativist

This type exists for example in Norway, Finland, Denmark and Germany, the basic principle is to ensure the participation of various interest groups in the supervisory and regulatory organs which is understood as a guarantee of diversity. In Germany the responsibility for broadcasting lies with the states of the Federal Republic, this results "in a uniquely decentralized broadcasting system with production centres in every region of the country". All broadcasting corporations are governed by an independent broadcasting council whose representatives are supposed to "reflect the 'socially relevant groups' of society". The representatives are either elected in the parliament or are delegated by various groups (including political parties, churches and labour organizations). Despite these provisions political parties have been able to gain influence in the Broadcasting Councils because "German parties are relatively strong in all segments of the political and social system and penetrate practically all of the 'socially relevant groups'". In the mid-1980s commercial competition challenged the public broadcasting system and a dual system was established. Private broadcasting is regulated by special licensing and supervisory institutions. Public service broadcasting is mainly financed by a monthly licence fee and advertising revenues (limited to twenty minutes each weekday).

Public Service

This type is characterized by a high degree of internal diversity, i.e. the needs of various interest groups are catered for by a large scale of programmes on the same channels. This system is typical, for example, of the United Kingdom of Great Britain and Northern Ireland.

The United Kingdom has a highly centralized communications system. Public service media are guided by the principles of high quality programming with a diversity of contents and general accessibility. Public service broadcasting in the United Kingdom tends to be central and national. Independent television companies run local television channels. The British Broadcasting Corporation is regulated by a Board of Governors, consisting of amateur regulators appointed by the government of the day. The BBC is required to commission twenty-five per cent of its programming from independent producers. It is financed by a licence fee and its Royal Charter has to be periodically renewed.

Commercial television was introduced in the United Kingdom in 1955 with the establishment of the Independent Television network. "In Britain commercial television adopted the traditions of a public broadcast service. It was modelled to redress weaknesses in the BBC, to make broadcasting more sensitive to popular taste, to promote regional culture and to oppose Londonism". In 1990 the duopolistic arrangements came to an end with the deregulation of the Independent Television network. All commercial television channels in the United Kingdom are regulated by the Independent Television Commission.

This brief description of the basic systems of public service broadcasting is not intended as a comprehensive overview but rather it was meant to illustrate that there are varieties within systems of public service broadcasting and that in each of them we can identify mechanisms that aim to ensure that the public is served in the best possible manner and that it is the public interest rather than organized partial interests (political, industrial etc.) that are served. The examples also illustrate that there are various ways of financing public service broadcasting (an issue to which I return later) and that public service broadcasters face competition from commercial broadcasters which has serious implications (not only in relation to the quality of broadcasting provided by commercial and public service broadcasters or the accessibility of broadcasting but also issues such as public aid).

There appears to be a prevalent notion that public service broadcasting is going through a crisis, or at least significant changes. Nicholas Garnham argues that "our inherited structures of public communication, those institutions within which we construct, distribute and consume symbolic forms, are undergoing a profound change". This change, he argues, is characterized by the ever-increasing power of the market, by a focus on television as an increasingly privatized, domestic mode of consumption, by the increasing gap between the information rich and the information poor and the shift from national to international markets in the informational and cultural spheres. John Keane (1991) argues that there are three principal reasons for the decline of public service broadcasting: fiscal squeeze, legitimacy problems and technological change. He argues that to treat the current public service media as "a bulwark of freedom against the confusions and limitations of commercial media" is highly problematic, just as it is problematic to think of public service media "as the paragon of quality, balance and universal accessibility....

The public service claim to representativness is a defence of virtual representation of a fictive whole, a resort to programming which simulates the actual opinions and tastes of some of those to whom it is directed". Nicholas Garnham argues that public service broadcasting is characterized by a "failure sufficiently to distinguish between two communicative functions within the public sphere: the collection and dissemination of information, and the provision of a forum for debate". Graham Murdock, draws attention to consequences for citizens when he writes that the current public broadcasting system failed to "keep pace with the proliferation of political and social discourses, has produced a crisis in the relations between public broadcasting and the viewer-as-citizen. This crisis in representation is exacerbated by the increasing tension between broadcasters, state agencies and government".

Stuart Hall identifies four key reasons for the crisis of public service broadcasting: technological, economic, political and social. The technological justification (spectrum scarcity) of the existence of public service broadcasting is no longer understood as valid by many, further, Hall points out that we live in an economic and political climate of a wholesale assault on "the very idea of a 'public sector'" amidst the growing social diversity of the audience (or in other words social fragmentation) and "the consequent pluralisation of cultural authority, which makes it increasingly difficult for broadcasters to see society as 'a public' at all or to speak to it as if it were still part of a homogeneous, unified national culture". Marc Raboy alerts that "problems of financing, mandate, and interpretations of purpose are all indications of a more fundamental problem of political will". He goes on to argue that in relation to the broader policy framework "the principal normative question will remain: What should be the public function of broadcasting in a democracy?". Garnham makes the significant point that "the public service, state-regulated model, whether publicly or privately funded, has in effect always been seen, not as a positive good but as an unfortunate necessity imposed by the technical limitations of frequency scarcity".

Carpignano et al. relate the crisis of public service broadcasting, among other factors, to "a crisis of legitimacy of news as a social institution in its role of dissemination of information about and interpretation of events (i.e., the social construction of public life)". James Curran gives a concise summary of Elihu Katz's arguments about the decline of television as public broadcasting.

Firstly, Katz suggests, the multiplication of television channels results in a fragmentation of the public thus "television has all but ceased to function as a shared public space. Except for occasional media events, the nation no longer gathers together". Secondly, civic communication is exchanged for high rating programs due to "the combined constraints of the new media technology, the new liberal mood, the economic and political burden of public broadcasting, and the seduction of multinational corporation" (ibid.). And finally, liberal democracy itself is endangered due to the growing separation between the television system and the nation state and the presumed weakening of national identities because it is in nation states where liberal democracy is practised and national identification is crucial in maintaining involvement in the democratic project.

Underlying this argument appears to be a notion that maintaining national unity is a desirable objective. Indeed, Paddy Scannell (1992) praises public service broadcasting for achieving this objective. Smith alerts us to the crucial role of media in the nation building project.

In looking at the role of the media in creating a certain uniformity within the nation-state, we are in essence looking at the process of nation-building, and at how the media are consciously brought into play to construct a "national" culture and a "national" community. Nation-states must have a measure of common culture and civic ideology, a set of common understandings and aspirations, sentiments and ideas, that bind the population together in their homeland.

Van den Bulck follows the same line of argument when stating that the system of public service broadcasting can be described as a typical and vital modern institution which played a crucial role in the modern process of nation building. She applies this to the case of Flemish public service broadcasting and demonstrates that it was given "the task of contributing to the creation and development of a national identity and culture. As such it had a threefold responsibility: education (as an extension of the national educational system), information (to create a political consciousness) and entertainment (to articulate a national culture)". In this respect attention needs to be drawn to Hall's identification of one of the threats to public service broadcasting: "on the basis of what cultural authority can a public service organisation speak to the nation?". He goes on to argue that broadcasting has a major

Perhaps the critical role – to play in this "re-imagining of the nation": not by seeking to reimpose a unity and homogeneity which has long since departed, but by becoming the open space, the "theatre" in this which cultural diversity is produced, displayed and represented, and the "forum" in which the terms of its associative life together are negotiated.... This cultural negotiation about the terms on which the centralised culture of the nation can be reconstituted on more openly pluralistic lines, remains broadcasting's key "public cultural" role – and one which cannot be sustained unless there is a public service idea and a system shaped in part by public service objectives to sustain it.

However, arguments relating to the role of public service media in nation building or the promotion of coherence should not only address the issue of inclusion in the mediated nation but also the equally crucial question of exclusion. The question of exclusion from public spheres is a central one as "public discursive arenas are among the most important and underrecognized sites in which social identities are constructed, deconstructed, and reconstructed".

My brief overview of literature on the crisis of public service broadcasting demonstrates that it has been approached from a number of angles and the sheer volume of literature on the subject is overwhelming. Nonetheless, James Curran adopts a more cautious attitude when discussing the decline of public service broadcasting. He rightly points out that it is difficult to establish whether arguments provided by those who herald the decline of public service broadcasting hold true because there exists "a misleading convention of equating public broadcasters with public service broadcasting. Public broadcasters' loss of audience is thus cited as evidence of systemic crisis. But, in fact, private broadcasters that are subject to effective public regulation are as much an integral part of the public service system, in the sense of serving welfare rather than purely market goals, as public broadcasters".

He identifies a serious lack in official proclamations about the role of public service broadcasting as "in none of these official documents is the facilitation of an open public dialogue or the extension of participation in collective discussion ever recognised as being an official objective of the public service system". In the official canon, Curran continues, "access" is defined as access to broadcast signals (an entitlement to reception rather than expression) and "diversity" is understood in terms of delivery (mainly invoked in relation to non-political programmes) and finally public affairs coverage has to demonstrate "due impartiality".

In this respect it is important to remind us that the establishment of public service broadcasting was not linked only to spectrum scarcity and an understanding of the spectrum as a public good but also to the view that "broadcasting should not be part of the new publicity system for consumerism but should provide resources for citizenship, should provide the kind of information, spaces for debate that people needed in order to participate fully in the new political process". In this respect it is worthwhile discussing issues of citizenship and broadcasting. Graham Murdock argues that there are three important ways in which the communications system is implicated in the constitution of citizenship:

1. The communications system provides access to information that is related to personal rights;
2. It enables access to information, interpretation and debate on areas that involve public political choices and
3. Citizens recognize themselves and their aspirations in the range of representations on offer within the central communications sectors and are enabled to contribute to the development and extension of these representations.

According to him public broadcasting played four key roles in organizing the new system of representation which aimed at extending citizenship rights. It enabled a public forum where platforms of the major political parties and legitimated interest groups were presented and packaged for consumption by the public at large. Further, it provided a new source of surveillance and feedback to those in power as well as (due to being a national service) creating associations between ideals of citizenship and definitions of the nation and its culture. It also redrew the boundary between the public and private spheres.

John Keane (1991) argues that public service broadcasting would serve its role better if there existed new constitutional settlements in all Western democracies that would make the exposure and repeal of the censorial methods of the contemporary state power possible. In more general terms Keane argues for government accountability to supra-national institutions such as the international civil society which is being formed with the help of new technologies. Keane, nonetheless, admits that new technologies are not necessarily heralding radical changes, a question to which I return in the last section of this chapter.

Moreover, he acknowledges that although nation states are undergoing changes, we cannot yet speak about their end thus we still need to deal with national rather than global public spheres. According to Keane the public service broadcasting system has to be radically redefined and opened up to a plurality of non-state media of communication. This process has to go together with a radical loosening of libel laws. A new constitutional settlement would also make it possible to achieve a "heterarchy" of media controlled neither by the state nor by commercial markets. For example, the BBC could remain funded by a licence fee but the present system of government appointments should be abolished. James Curran (1998) points out similar issues when arguing for a rethinking of the rationale of public service broadcasting and the reform of its actual practice and organization. In addition, Curran argues, there is a need for a revision of the objectives of public service broadcasting in Britain which would foster a change in its style of journalism. "This is still profoundly influenced by a civil service/professional model which stresses the disinterested mediation of information, the imparting of knowledge and the impartial umpiring of differences of legitimate opinion.

It is a mandarin-like conception in which the electorate, the rulers of democracy, are briefed by intelligent and responsible public servants rather than merely entertained by market spectacle". This change is especially important as "an over-great stress on legitimated forms of public knowledge and accredited speakers unduly restricts participation in this dialogue. Indeed, this is a constantly repeated refrain of much academic research which suggests that TV news and current affairs is often defined by elite assumptions and sources". There is, nonetheless, Curran suggests, a reform movement within the broadcasting community "which is intent upon extending social access and expanding the range of voices and views on air". This reform movement manifests itself in the form of new phone-in programs, audience participation formats and access slots.

Other authors argue that new genres provide a different way of understanding the role of public in media. Carpignano et al. argue that talk shows are the only program on which public as appearing on television gets full recognition. They propose that it is the only genre that proclaims its intention of presenting issues for public debate, moreover it is the only program whose discursive format is conversation. This points to a change in the understanding of the audience, studio audience is participating not only in the viewing but also in the actual "writing of the script".

TECHNOLOGICAL CHANGE

The above overview has already hinted at the importance of technological changes in relation to broadcasting as such (it has been argued that spectrum scarcity is no longer a valid justification of the existence of public service broadcasting, new technologies enable more interactivity etc.) and also in relation to the possibility of creating arenas for public discussions. In relation to the possibility of creating an overarching space for public discussion the Internet immediately comes to mind. A number of politicians understand new technologies as playing a central role in bringing about social and political change (not least a reinvigoration of democracy via e-government). In this respect it is enough to think of Al Gore and Hillary Clinton both of whom envision the World Wide Web as a means for a global conversation that would bring about fundamental change.

However, as Robins and Webster (1999) point out this argument is flawed as it presents social problems and inequalities in contemporary societies exclusively in terms of a failure to communicate. Murdock and Golding argue that the Internet appealed to politicians because "as a solution to the problems generated by the accelerating dynamics of marketization and the decline of public welfare systems, it offered several advantages. It was relatively inexpensive in terms of the public investment required, it offered scope for partnerships with private companies, and it could be presented as a creative and forward-looking response to the inevitability of tec hnologically driven change."

John Keane (1991) as well as Tod Gitlin (1998) argue that the World Wide Web is possibly providing the means for the creation of an international civil society and increasing the possibilities for a plurality of publics. Yet, as Gitlin points out, "there is one problem which the new means of communication do not address and may even worsen: the existence of a two-tier society. To those who are information-rich (or information-glutted) shall more information be given". Murdock (2004) draws attention to the fact that exclusion from the Internet is not only the result of the financial costs involved with being "connected" but also to "feelings of incompetence, symbolic exclusion, and the irrelevance of what is currently on offer".

The World Wide Web has been seen as an open, fluid and flexible space that makes direct and immediate contact possible thus facilitating new forms of community in which the basis of social conflict and for that matter difference are overcome.

As far as the Third Way politicians are concerned, virtual technologies promise to make community instantly available, now as a service (or a commodity) that can be piped into the electronic home - community, or interactivity, for domestic consumption.... The politics of connexity seeks to work in harmony with this fortunate and benign logic, "breaking down the barriers and separate identities that have been the main cause of human suffering and war". The objective is to promote a greater sense of transparency, mutualism and trust- the immediate relations of face-to-face community are a key reference point (we are even told that new technologies promise "direct connection between people's minds, transcending the idea of separate selves and subjects").

A number of authors have written critically in relation to the utopian notions connected with the Internet. For example, Graham Murdock (2004) points out three problems with the Internet. First of all, access to it remains highly stratified (in terms of income, age and education). Secondly, it segments its audience, taken together these technologies make it entirely possible to only watch what one already enjoys and to only encounter opinions one already agrees with. In a situation where worldviews are increasingly polarised and talking across differences on a basis of knowledge and respect is more vital than ever to a working deliberative system, this hollowing out of collective space presents a major challenge to democratic culture.

And finally, it (just like any other branch of the culture industries) became an arena for corporate activity. Frank Webster points out that "what we have here [in the utopian visions of the Internet] is the assumption that quantitative increases transform – in unspecified ways – into qualitative changes in the social system." It does not come as a surprise that Habermas remains sceptical of the potential embodied in the Internet:

Whereas the growth of systems and networks multiplies possible contacts and exchanges of information, it does not lead per se to the expansion of an intersubjectively shared world and to the discursive interweaving of conceptions of relevance, themes, and contradictions from which political public spheres arise. The consciousness of planning, communicating and acting subjects seems to have simultaneously expanded and fragmented. The publics produced by the Internet remain closed off from one another like global villages. For the present it remains unclear whether an expanding public consciousness, though centered in the lifeworld, nevertheless has the ability to span systematically differentiated contexts, or whether the systemic processes, having become independent, have long since severed their ties with all contexts produced by political communication.

In relation to broadcasting the most significant development is the advent of digitalisation. According to a dictionary entry digital communication involves "a process whereby the transmission of information – letters and numbers, voice, facsimile or video – is coded into discrete on/off electronic signals, in contrast to analogue transmission in which a signal is a measure of time of a continuous flow of electricity." To explain this very rigid definition I outline the implications of digitalisation for television. Digital broadcasting involves the transmission of digital signals to a digital television set, these signals can be broadcast over the air or via cable/satellite. "With its brilliant, high-definition images, CD-quality audio, and the possibility of transmitting multiple programs and information simultaneously, the quality of digital television is a vast improvement over the analogue television that most of us experience today".

Digitalisation enables the transmission of more visual information (usually an analogue picture has 480 vertical interlaced lines with about 340 horizontal pixels per line, a digital television picture has 1,080 vertical interlaced lines with 1,920 horizontal pixels per line) without increasing the broadcast frequency spectrum as data is in a compressed form (the frequency that carried one analogue television channel can carry at least four digital television services and radio and text services.). The introduction of digital television appears to be a matter of the next few years and public service broadcasters are expected to lead the way in its implementation. Problems that occur with the implementation of the switchover include huge costs (e.g.

Jones 2003 gives an assessment of US system-wide costs of 1.6 billion USD), these, however, involve consumers as well (the need to buy set-top boxes), additional problems include the threat of piracy as well as control over gateways. Mosco draws attention to the fact that the development of technical standards itself presents significant problems as digitalisation requires the harmonization of the production, distribution and reception of digital signals. In this respect a telling example is that of the UK. In March 2005 the first village – that of Ferryside in Wales, UK – switched to digital. In the UK more than a million set-top boxes were sold over Christmas 2004 bringing their price down from about 100 GBP to 35 – 40 GBP. Yet, the Digital TV Group (which was set up to promote digital television) admitted that "almost half of the transmitters needed to send signals to Freeview set-top boxes are still using old technology and will not be fully upgraded until 2012". This means that the quality of pictures on television sets linked to the set-top boxes is poor and it is likely that customers facing this problem will have to spend 300 GBP or more on new aerials and cabling.

On the surface (and despite the above mentioned problems) digital broadcasting seems to be a blessing: it will result in the multiplication of the number of available channels, better quality products, more competition etc. Yet, the initial excitement that surrounded the promise of digital television and in particular of digital terrestrial television (DTT) did not match actual developments. By 2002 DTT failed in Great Britain (with the collapse of ITV Digital) which seriously jeopardized the British government's ambition to make Britain the first fully digital television nation by 2010. A closer look at digitalisation clarifies that other issues are at stake as well and that digitalisation certainly does not make public service broadcasting and government regulation in the sphere of broadcasting obsolete, rather the contrary. In an influential public policy report from 1997 Graham and Davies convincingly argue that the introduction of new broadcasting technology will not significantly increase competition and broadcasting will remain highly concentrated.

Broadcasting is an example of an economy of scale (characterized by high fixed costs and very low marginal costs), with the introduction of digital broadcasting it will turn into an economy of scope (activities in one area decrease costs respectively increase revenues in another area). They alert to the possible monopoly control over gateways as well and argue that with the new technology "under 'free market' conditions consumers will face a choice between a narrower range of cheaper (and yet still high quality) broadcasting and a broader range of more expensive and yet lower quality programmes". The introduction of digitalization will not prevent the continuing fragmentation of the audience, in fact, Graham and Davies argue, this could be achieved by public service broadcasters which is another reason for maintaining (and in fact improving) their services. Murdock (2004) argues in a similar vein: "Public broadcasting also has the capacity to counter fragmentation. Employing programmes, whether watched in real -time or retrieved on-line, to kick-start on-line activities maintains at least a minimal base of shared experience. In addition, web surfing has the capacity to counter the self enclosure of zapping and personalised video recorders. Although mixed programming now often takes second place to crafting distinctive channel identities, a user entering the broadcaster's web site to pursue a particular interest will encounter a wide range of other possibilities, some of which they may be tempted to follow up."

Graham Murdock (2004) concentrates on the active and positive role that public service broadcasters can play in the creation of a new cultural space, a "digital commons, that can help forge new communal connections and stand against the continual pressure for enclosure coming from commercial interests on the one hand and the new moral essentialism on the other." However, developing these resources requires us to abandon our old analogue maps of the cultural industries which depicted a series of stand alone institutions separated by incompatible technologies and compile a digital chart showing public broadcasting as the central node in a new network of public and civil institutions that together make up the digital commons, a linked space defined by its shared refusal of commercial enclosure and its commitment to free and universal access, reciprocity, and collaborative activity. This space is potentially global in scope.

This chapter began with an evaluation of Habermas' concept of the public sphere which exists (existed) primarily as a discursive space with a distinctly critical role. Despite the various criticisms related to Habermas' seminal work there are at least three features that are of key importance when assessing public spheres in contemporary societies. Namely: the link between mass media and democracy; material resources required for the public sphere and last but not least the fact that media in their critical role are threatened by both the market as well as the state.

Although Habermas argues that media in contemporary societies create only a pseudo public sphere and arguments have been proposed in relation to the existence of multiple public spheres I focus on public service media (more concretely on public service broadcasting) and the legislative framework related to it to analyze whether legal instruments and regulatory mechanisms reflect any of the above mentioned three dimensions of Habermas' concept, i.e. public service broadcasting has a critical role to play in relationship to democracy, it must be guaranteed a material base for its existence and be independent from both the state and the market.

The reason why public service broadcasting (rather than private one) is crucial for my argument is that it carries a certain "public sphere ethos" due to its very remit and despite the well-documented fragmentation of the audience it has at its core the promise of reaching the most numerous and most diverse audience. Also, public service broadcasting, unlike private media, importantly addresses its audience as citizens – as a polity – rather than purely as consumers. Examples of concrete systems of public service broadcasting illustrate diverse ways of implementing goals that are basically intrinsic to the concept of public service broadcasting (i.e. universal service, accountability, public funding (at least partial), regulation of content and a limited number of service providers).

The chapter has also dealt with the presumed crisis of public service broadcasting, even those who object to the term crisis acknowledge that the last decade has brought a serious questioning (unfortunately often not followed by a public debate) on the role of public service broadcasting in contemporary democracies. This questioning is closely related to the forms and justifications which mark the origins of public service broadcasting (albeit these, as I argued, are often misrepresented and misunderstood) and is also directly linked with a general deregulatory ethos and the emergence of new technologies (digitalization is of major importance in this respect). In particular technological developments have been surrounded by much (unfounded) optimism and hype. From a regulatory point of view the argument that new technologies justify deregulation is utterly unfounded. On the contrary, as I suggested in this chapter and argue in the following ones, there is in fact need for more regulatory rigour. Although digitalization is not proving to be the universal solution that many expected it to be, it certainly provides avenues for the rejuvenation of public service broadcasting – this, though is a huge challenge and, as I argue further, it is not at the least considered by regulators in the cases I deal with.

This chapter in essence provided a probe into the academic debate surrounding public service broadcasting and its roles (among them the creation of a critical discursive space), however, the world of the academia is often miles away from the world of legislators and political elites. My aim in the following chapters is to confront the visions and solid arguments outlined in this chapter with concrete legislative measures related to public service broadcasting in the European Union, the Czech Republic, the Slovak Republic and the Republic of Ireland. Before moving on to the case studies, however, the following chapter considers the general regulatory framework and the most important developments within it in the last fifteen years.

RE-MODELLING THE PUBLIC SPHERE: THE CASE OF TELEVISION

One of the most elaborate adaptations by media scholars of the Habermasian public sphere for a specific thematic is Peter Dahlgren's book Television and the Public Sphere. Citizenship, Democracy and the Media (1995), recently refined by his update article Television, Public Spheres and Civic Cultures (2005). Dahlgren addresses the public sphere idea in the context of television by integrating its various readings and critique, by crystallizing main themes and questions that thus arise, and by mapping the issues under the categories of (1) social structures; (3) media institutions; (3) sociocultural interaction; and (4) media representations. Through this discussion, Dahlgren thus comes up with a conceptual framework, in which

"[e]ach dimension serves as an entry port to sets of issues about the public sphere, both theoretical and conceptual questions as well as empirical and evaluative ones about its actual functioning.

I find Dahlgren's basic approach useful as it provides a structured way to theorise as well as empirically analyse the public sphere. Arguably, as Dahlgren notes, the four dimensions are closely intertwined and condition each other; yet, the framework makes it possible to conduct focused and systematic empirical analysis on a specific dimension. Yet another strength of Dahlgren's reworking of the public sphere concept is his update (2005), in which he takes into account of the developments of the past decade. He notes that the technological developments lead to technological convergence between television as we know it and other media, but even more importantly, he argues for integration of television and other forms of media culture. He also brings forth the idea of multiple public spheres and emphasizes the increasingly blurring boundaries of the public and the private.

He recognizes the importance of popular culture in television and advocates utilization and combination of the 'culturalists' (Dahlgren's term for cultural studies oriented) perspectives with the more traditional social sciences oriented views to the thematics television and the public sphere; this is a parallel to his ideas of interlinking political science, public sphere and culturalist approaches.

However, Dahlgren's (2005) most interesting elaboration – at least regarding the linkage between empirical analyses and a broader perspective of the relevance on television culture – is that instead of concentrating on citizenship (as in 1995) he stresses the notion of civic cultures. He re-emphasises the fact that television's role is not only to provide a sphere for expressing such cultures, but also in pre-structuring them.

He also lays out a conceptual model of five dimensions of civic cultures:

1. Knowledge and competence,
2. Values;
3. Affinity and trust;
4. Practices; and
5. Identities.

They can, according to Dahlgren, be seen 'as integrated circuits with five mutually reciprocal dimensions' and that can be taken 'as starting points for empirical inquiry about the media's significance for civic cultures, and (...) on the role of television'.

THE FOUR DIMENSIONS OF THE PUBLIC SPHERE

Dahlgren defines the four basic dimensions of the public sphere as follows:

1. *Social structures*: Dahlgren formulates this category constituting of institutional arrangements of society, including social stratification, power alignments, and the state. Thus, it entails political, economic and legal aspects, as well as, for instance, the nature and quality of educational system. Dahlgren admits that this dimension is potentially so broad that at some point of the analysis it must 'be put in brackets', in order to be able to remain the focus in the public sphere (op cit. 12). However, he strongly emphasises Habermas' point of the public sphere as historical, connected to and constructed by larger social figurations. Therefore, although social structures can be said to constitute and condition the other aspects of the public sphere – in his words, institutions, interaction and representations – it is important to recognize this overall dimension. In this context, Dahlgren raises an issue that reflects to the complex conditions of today's societies and that he calls 'the tension between the whole and the parts': should there be a single (if pluralistic) public sphere or can many spheres together constitute one?
2. *Media institutions*: By this, Dahlgren refers to the organization, financing and legal aspects – from those governing ownership and control to those pertaining to constrains and freedoms of communication. He brings up the issue of (national) public service broadcasting in the commercialising and globalizing television markets as one of the central themes of this dimension.
3. *Sociocultural interaction*: This dimension 'refers to non-mediated face-to-face encounters between citizens, to relevant aspects of subjectivity and identity processes and to the interface of media and citizens' (op cit. 12). Dahlgren sees the last point, that is, processes related to media reception, being often the first step of interaction and notes:
   - "[T]he public sphere is not just a 'marketplace of ideas' or an 'information exchange depot', but also a major societal mechanism for the production and circulation of culture, which frames and gives meaning to our identities.".

   Accordingly, the question of interest in this dimension is if and how people's identities as citizens and/or consumers are constructed.
4. *Media contents*: Dahlgren uses the term media representation, and defines it as concerning what is portrayed, how topics are presented, what modes of discourse are at work and what the character of debates and discussions are. However, I suggest the term media contents, as it can then include television output both with very broad, overall quantitative terms, while also addressing specific

'representations'. Dahlgren notes that this dimension simply points to the fundamental questions: What should be selected for portrayal and how should it be portrayed? This dimension, then, calls for monitoring of the media, for critical analyses of infinitive perspectives.

The internal dynamics between the various dimensions of this public sphere model, although not explicitly discussed by Dahlgren (1995), are sketched. The figure is intended to illustrate the encompassing nature of social structures that shape institutions, social interaction as well as contents. This is surely a two-way dynamics, but the 'feedback' from the three is not as overt or pronounced. I also suggest the self-evident notion that media contents are central to the idea of television as a public sphere. Similarly, as darker and lighter arrows indicate, institutional setting (e.g., whether a broadcaster is a public service or commercial one) tends to influence, in general terms, contents more strongly than vice versa. The dynamics between media contents and social interaction should and could ultimately be free flowing, media output fostering interaction and interaction feeding into media contents. The basic 'dumbing down' argument seems to insinuate that this relationship does not work sufficiently well from the contents to interaction and thus democracy is not fostered; and the critique of the public sphere model reminds us of the fact that the opposite flow is often very homogenous, that is, the multitude of possible voices are not reflected in the contents.

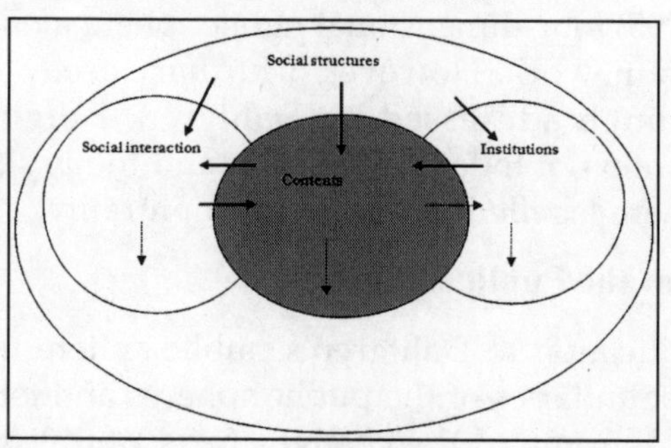

Fig. Four Dimensions of Television as a Public Sphere

Dahlgren's (1995) model, in sum, raises several issues. One pertains to the question to what extent television in fact can constitute a public sphere in today's diverse societies. More elaborately, in Dahlgren's words: What is the relationship between the rational/analytic versus affective/aesthetic modes of communication, and can the typical analytical connection between politics and rationality versus entertainment and emotionality be contested? Dahlgren recognises the value of Habermas' theorisation of the public sphere, but argues that by no means it is a concept that should remain unchanged. He positions the public sphere as a historically contingent, negotiated and contested space that is constructed also in the 'discursive practices of civil society'.

Furthermore, Dahlgren's four dimensional model (1995) recognises the political economy aspects and commercial imperatives of television, as well as viewers' various positions, but neither glorifies popular culture nor emphasises any normative evaluations. Instead, he points out particular features of television (journalism) that defy the traditional binary divide of 'public knowledge' and 'popular culture'. Thus, Dahlgren's framework allows for an approach that can be distanced from the often value-laden commentary based on traditional opposites. At the same time, this framework invites and requires discussion on the extent, potential and fluidity of television's role as a public sphere – or, for instance, the possible existence of a 'popular' or 'oppositional' public sphere – as, according to Dahlgren,

"Television operates in this late modern setting as an industry, as an incessant producer of audio-visual discourses which have a central position in the semiotic environment. As an industry, television has to follow the precepts of audience maximization and profits; moreover, it is the paramount vehicle of consumer culture. While television is the dominant medium of the public sphere, 'public sphering' is clearly not television's dominant purpose, and its institutional logic greatly conditions its role within the public sphere. As a sociocultural experience, television provides symbolic 'raw materials' which are experienced and reflected upon in varying degrees."

SPECIFYING PUBLIC SPHERES: FLUIDITY AND PLURALITY

Dahlgren's (1995) four-dimensional public sphere model as such provides for a theoretical framework. However, I further specify it by two thematic notions that, although addressed by Dahlgren, I argue deserve further emphasis in the context of today's television and today's (western) societies. The concepts are here labelled as fluidity and plurality.

Fluid Boundaries of the Public Sphere

The first specification to Dahlgren's public sphere model concerns the definitions of the boundaries of the public spheres and television's historical position. To address some of the changes for a pluralist, descriptive, non-normative framework for television's potential in the formation of one or more public spheres, I add an attribute to Dahlgren's model that deals with recent theorising of changes in overall societal conditions, the public sphere in relation to other societal spheres, and main changes regarding television as a medium. I borrow the basic term 'fluid' from Zygmunt Bauman's work on Liquid Modernity (2000).

While considered as a key theorist of postmodernism, Bauman seems to imply in his recent work that we live in a world in which some characteristics are still fundamentally modern, yet some are in a process of constant change. Therefore, with the term 'fluidity', I wish to describe the present social conditions that in my view surround television's role as a potential public sphere. In Bauman's latest writings, he has indeed replaced the term postmodern with the idea of modernity as liquid or fluid. With this shift, he seeks to emphasise that we are not past modernity, but in a transitional stage where old characteristics of the society mix with the new. 'Fluid' as a metaphor, then, refers to a situation where society's key modern characteristics – the pursue for stability and permanence that equal safety – have faded. Instead, we are heading towards a society dominated by insecurity, unsafety, and uncertainty. These, in turn, are caused by ever changing, liquid state of affairs. The central features of such a society are, for instance, the importance of time (as everything is only of momentary character); the mobilization of power that becomes more and more slippery, and harder to track down; and the disintegration of social networks. The disintegration means disengagement from common frames of reference and any kind of dependence – for example, the growth of differentiation and individualization, and as a result, the disappearance of traditional Politics 'with a capital P'. Instead, people engage in individual life politics that concentrate on a lonely life of achieving and consuming.

In this context, Bauman addresses the media, television in particular and claims that the public space or sphere (terms used in his work synonymously) is becoming empty. This is not an original idea as such but seems to be in line with the main strand of the tabloidization discussion. However, Bauman argues that this shift is not abstractly caused by the Zeitgeist of the times, but is a result of a fundamental transformation of the public sphere. He gives it a new provocative definition: 'public space is where public confession of private secrets and intimacies is made' (op cit., 40). Yet, he does not remain with the criticism of consumerism and entertainment-orientation. Bauman sees a more fundamental meaning in the lightening of media output.

He claims that the question is not about the ever shifting boundary between what is private and what is public, but that in the public sphere, citizenship-based politics have been replaced with public display of private affairs. The media, then, do not act accordingly only out of, for instance, commercial imperative to entertain, but they, television in particular, feed into a specific need of the liquid society. The media provide individual solutions to shared problems: they create momentary spectacles so that people can pretend they belong, and that they can see how other lonely individuals cope with individual problems of life politics:

"Each day the first-page press and first-minute TV headlines wave a new banner under which to gather and march (virtual) shoulder to (virtual) shoulder. They offer a virtual 'common purpose' around which virtual communities may entwine … [T]hey effectively ward off the condensation of 'genuine' (that is, comprehensive and ever lasting) communities which they mime …They scatter instead of condense the untapped energy of sociality impulses and so contribute to the perpetuation of the solitude…" (op cit., 201).

In addition to the strong criticism on the media of the 'liquid modernity', Bauman also has a solution in the form of a strong view on the notion of the public sphere. In his view, the societies defined by the 'unholy trinity' of uncertainty, insecurity and unsafety require joint action and some forms of consensus (even at a global level) instead of individual discursive identity construction and private resistance of the power-block through consumption:

"Any true liberation calls today for more, not less, of the public sphere and public power. It is now the public sphere which badly needs defence against the invading private – though, paradoxically, in order to enhance, not cut down, individual liberty" (op cit. 51).

By utilizing the concept 'liquid modernity' as a metaphor surrounding the social structures as well as other dimensions of the public sphere, I seek to clarify following theoretical issues: First, Bauman's argumentation helps to refuse the term of postmodernism and as characterising the social conditions of today. While I see much of the postmodern mode of media theorising useful, especially regarding popular television culture, my choice of the public sphere paradigm points to another basis of the study. Accordingly, I agree with Preston's interpretation on Habermas' opposition on postmodern theorising in reference to both techno-utopian rhetoric regarding media technologies as well as to the most optimistic strands of cultural studies referring to ever-resisting active audiences. Both views seem to carry features of neoliberal thinking that advocates separation of common, traditional politics from moral-practical justification.

Also, while Bauman denies the optimistic postmodern views on the media, he also can be interpreted to refuse the idea of complete refeudalization of the public sphere. Rather, Bauman's strong views on how privatization of the public sphere is connected to the conditions of liquid modernity re-iterate a normative presumption or an ethical basis of the public sphere model, but in the context of today's societies. Whether a local or a global scale is in question, the underlying principle of the modification of the public sphere model is that the media should participate in forming a democratic public sphere (if not many), by giving access and representing different viewpoints (or, in Dahlgren's terms, 'civic cultures'), as well as by assisting in formulating and realising society's common goals.

Public and Other Spheres

Yet another consequence of emphasising the fluid nature of modernity helps to set boundaries to the public sphere as an analytical concept and to position Dahlgren's four-dimensional model of television as a public sphere in relation to other spheres in the society. By acknowledging the possibility of fluid modernity, one acknowledges rapidly changing structures the current societies and the fluidity of various boundaries. For this purpose, I find Thomas Janoski's model of four societal spheres – the state sphere, public sphere, market sphere and private sphere – useful. In his model, the state sphere includes the executive and juridiciary powers, bureaucracy, as well as the police, military, and the like. The market sphere consists of different markets, and firms operating in those markets, as well as labour unions influencing the sphere. The private sphere is that of family and friends, of intimate relationships.

Janoski's (1998) understanding of the public sphere, then entails voluntary associations, social movements, self-help groups, education and health care – as well as, essentially, the media. However, he emphasises the fact that the borders between the spheres are by no means fixed and strict, but that the spheres overlap. Accordingly, he positions television not only as a component of the 'pure' public sphere. One overlapping area is between the public and state spheres, in the realm in which he also positions political parties and which for him is the realm of public service broadcasting. The other area where television is present is situated in the meeting of the public and the private spheres: this occurs when people's private lives disclosed in the media.

Janoski himself recognizes the much discussed blurring boundaries of the public and the private spheres and states that increasingly in modern times, the state, market, and the public spheres have invaded the private sphere. Another increasingly fluid border or overlap could be seen between the market, the state and the public sphere, as the economization, or the distance between the economical and the political have been said to diminish if not vanish.

In addition, many scholars who address the so called marketization of the media would most likely like to place for instance much of today's television in an intersection of public and market spheres. Also, while Janoski's nation-based model is relevant to the examination of television with a nation-wide focus, internationalisation and globalisation tendencies influence the spheres. Indeed, it has been argued that globalized media markets have already emerged full force, and the medium most affected is television; both regarding ownership as well as contents. With these alterations in mind, I argue that Janoski's conceptualisation of the location of the public sphere helps to understand the dynamics and overlaps between various societal spheres.

Television in the Era of Availability and Plenty

The final additional emphasis regarding fluidity concerns a changing characteristic of television as a medium. As Dahlgren (2000) mentions, recent changes specifically influencing television in Europe include deregulation, diminishing support for public service broadcasting and a shift towards neoliberal attitudes on media's role in the society, as well as a general individualization of culture. However, I borrow a further theoretical notion of television's historical position, complementing Dahlgren's model, from John Ellis' work Seeing Things. Television in the Age of Uncertainty (2000). Based on above mentioned developments, and taking technological advancement into account, Ellis traces television history as having developed from the 'Era of Scarcity' and the following 'Era of Availability' to the current 'Era of Plenty'. He is inspired by Bauman's theorisation of the current social conditions and implies that the Era of Plenty coincides with the 'Age of Uncertainty', or, in other words, in the conditions of liquid modernity. Ellis (op cit., 72) argues that in the times of fragmented audiences and increasing choice of television contents, this medium can provide people with a process of 'working through' the complex nature of our liquid societies. He refers to Bauman, yet takes a more positive stand than the latter, in emphasising television's importance to the Age of Uncertainty: "[C]old tolerance is not enough between differing beliefs and the lifestyles that accompany them. What is needed is 'sympathy', and sympathy is the product of television's process of working through. By placing explanations, rearranging the facts, looking from all possible angles, using the different emotional registers of its different genres, television is able to provide sympathy across the process of differentiation that is modern consumer society." (Op cit., 85)

The relevance of Ellis' thinking to the public sphere model used here is that television is taking increasingly different institutional forms, providing more contents with new technologies, potentially involving diverse social structures and social interaction in diverse ways in different situations – thus contributing to its fluidity as a public sphere. By the same token, as Ellis notes, the working through is an open-ended process without a conclusion, but television's power is in its ability to provide for different sense-making processes that contribute to understanding of the uncertain times of liquid modernity. Dahlgren, too, recognizes Ellis' orientation as important in theorizing television and the public sphere, and suggests three expansions to it, mirroring the idea of fluidity: First, he notes that there are multiple trajectories such as class, gender, ethnicity, but also technology as well as local, national and global settings – and they all shape working through in complex and contradictory, un-fixed ways. Second, working through is different for different kinds of groups, and thus television's 'hegemony is loose, leaky and always at risk' (418). Third, Dahlgren wants to emphasise that Ellis' approach is not polarized between affective and rational – in other words, working through could be said to be a fluid process between the two.

In sum, to include definitions of historical specificity that emphasise the fluid nature of television as a public sphere to Dahlgren's model, I want to:

- Frame it within today's (western) world as fluid modernity,
- Locate it in shifting, overlapping relations with other societal spheres, and
- Recognise that television as a medium has entered the Era of Plenty that may contribute to the model of television as a public sphere that is changing - not towards a specific direction, but fluidly.

Plurality, Viewers' Positions and Civic Cultures

The second specifying concept to Dahlgren's (1995) model of the public sphere, plurality, is indeed recognised by him. Yet, I discuss the concept further, as it is at the core of Dahlgren's reworking of the public sphere thematic (2005), and as it has been widely discussed in relation to the classic Habermasian public sphere in many contexts. I find it paramount to overcome the dual definition of the position of the viewer as viewer-citizen (as it is central in the original Habermasian public sphere framework) and the viewer-consumer (as the view is emphasised by the postmodern cultural theory-based of research). This, in turn, leads to the fundamental re-definition of the public sphere idea as containing plurality in itself, and specifically, to the view that television, in Dahlgren's (2005) terms, fosters civic cultures – in plural.

At the outset, based on my reading of Dahlgren's (1995) four-dimensional public sphere model, I define mass communication, including television, as mediated communication that potentially reaches mass audiences. Yet, as many theories as well as empirical evidence have proven, the masses are in today's communicative context seldom reached; instead, current overview in the European situation reiterates the 'fragmentation' or 'polarization' of television views as relatively fluid groups of audiences. This means that viewers seem to form all smaller subgroups or, in business terms, 'audience segments', or the viewers divide as audiences of certain few genres. Also, an audience consisting of a group of viewers is not synonymous to a public. In contrast to the term 'public' as accessible to all, used for instance in the case of public service broadcasting, Habermas notes that the most commonly used meaning of 'the public' refers to the public carrier of public opinion as a 'critical judge' to counter official powers. Admittedly, public service broadcasting ethos often seems to imply that the condition of an institution being 'public' best serves the formation of 'a public' of 'the public'. Based on Dahlgren's (1995) theorizing, I see the concept of 'a public' as a 'critical judge' as a potential, neither a normative ideal nor a descriptive analytical tool.

Furthermore, the concept of television viewers as citizens has undeniably been at the core in thinking inspired by the public sphere paradigm. The citizen entails the modernist idea of 'passive and active membership of individuals in a nation-state with certain rights and obligations'. In contrast, as already mentioned, (postmodernist) cultural theory has often been considered to address the viewer as a consumer, if an actively resistant and reflective one. Toby Miller states, somewhat ironically, that when discussing television, there still exists a complicated relationship between citizen and consumer, 'even though the citizen is a wizened figure from the past and the consumer is a naïve phenomenon, essentially a creature of the nineteenth century'.

However, Dahlgren's model clearly indicates that television may provide positions of the citizen or the consumer in different times, or combine them. Even further, television audiences may be offered positions beyond the dichotomies, such as the participant or the player, as for example Trine Syvertsen (2004) suggests. Indeed, there seem to be competing conceptions both in the academic discourses as well as on public discussion on public communication and the kinds of collective groups it forms, summarized and labeled aptly by Paschal Preston's as Civic National, Affective National, Postmodern Cultural and Global Information.

The Civic National idea is based on citizenship, and here, in a true Habermasian fashion, identity means rational, critical members of a political community that collectively form a public sphere. Affective National, in contrast, is according to Preston's model a combination of a political identity and particular cultural or ethnic identities. These various identities take distinct national and other civic and social formations. The key actors, in this view, are citizens and bearers of a distinctive cultural or expressive identity. In contrast, the predominant identities of the postmodernist cultural view are that of active audiences, in a situation where national and other modern social identities have collapsed. These identities, then, are individual and multiple, and the key actors are active audiences or reflexive consumers; sometimes defined as 'socially situated' audiences. Finally, the global information society view (in its neo-liberal version) sees the key actors of public communication simply as consumers of media products in the market-place.

Surely, as Preston (2001) emphasizes, the above are abstract analytical categories that will be combined in many 'discursive accounts and practices'. But exactly as such they highlight that in the blurring world of the public and the private, citizenship is perhaps no longer the sole sufficient concept, even when theorization stems from the public sphere tradition. Plurality is at stake here, too. Also, television viewers do not engage a systematic and static use of television as a public sphere, and while the critical-rational citizen may be the key participant in a democracy, it may not be the only or the 'pure' position for participating in a public sphere. For example Ellis (2000) uses the concept of witness; and in my view his idea implies that shared witnessing and 'working through' with television may be a part of a public sphere, alongside with more active and explicitly rational debate and opinion formation.

However, I want to emphasize that I do not wish to replicate what Preston (2001) calls the 'Postmodernist Cultural' stand, in that the existence of 'modernist identities' would be denied and that the communal, shared, unifying elements of public communication for the traditionally political would not be recognized. This simply implies that televised public communication is not for citizen-viewers or consumer-viewers only. By buying into one particular conception would mean to undermine the pluralist nature of the medium. As Dahlgren (1995) states:

"Television fragments and unites, it clarifies and occludes, it informs and distracts. (...) In terms of television's role in the public sphere, we would be foolish to exaggerate its contribution to critical reason and progressive affect

(...) Dahlgren (2005) has taken this exact issue as one of his main revisions of his public sphere model and defines his view on citizenship in today's societies through the notion of political agency, as one mode of individual and collective action. Therefore, he emphasize the idea of identity formation (citizenship being one identity or rather, a part of an identity of an individual) and then goes on launching the notion of civic cultures, 'as referring schematically to those cultural patterns in which identities of citizenship, and the foundations of civic action, are embedded'. He parallels civic cultures to Putnam's (2000) idea of 'social capital' in that civic cultures are resources for individuals and groups when they act as citizens.

In further mapping such resources and television's role in enhancing the resources, Dahlgren openly declares his pluralistic view in the theoretical realm. He wishes to overcome the main flaws and utilize the main strengths of three traditions. Thus, he not only draws from the public sphere perspective but couples it with political science – in order to keep the formal democratic politics in the picture. In addition, he turns to culturalist approaches that help in understanding of processes of the production of meanings. With these tools at hand, he constructs the five circuits of civic cultures that, while integrated, can 'be individually thematized and studies in empirical cases'.

The dimensions and their relation to the public sphere disposition are summarized below:

- *Knowledge and competence*: With this circuit, Dahlgren stresses the current existence of multiple, fragmented public spheres that possess different knowledge and competencies; but also the problematic nature of such a situation for a functioning democracy.
- *Values*: While Dahlgren notes that values enhancing democracy can be disputed, he clearly points out the possibility of the media (television), as 'civic cultures need to underscore commitment to both rules of the game [procedural values of a democracy] and the larger visions of a democratic society'.
- *Affinity and trust*: In Dahlgren's words, affinity and trust point out to the need for 'a minimal sense of commonality among citizens in heterogeneous late modern societies'.
- *Practices*: This circuit deals with the notion that a democracy is built on and functions because of certain practices, discussion and rational decision-making being the traditional practices connected to the public sphere ideal. Dahlgren extends this circuit also to entail, e.g., practices of television viewing – that may be said to relate to individual control and empowerment.
- *Identities*: This circuit is explicitly concerned with the identity formation of the viewers, and the ways they are addressed. While citizen is not the only identity, it is still needed to foster a democracy. Thus, citizenship in turn still needs to be fostered for example by the media.

To condense the model, Dahlgren's (2005) five dimensions of civic cultures, although thematically distinct, clearly intertwine. Each of them can be supported, renewed and re-worked in a mediated public sphere, and they, in turn, provide resources for citizenship.

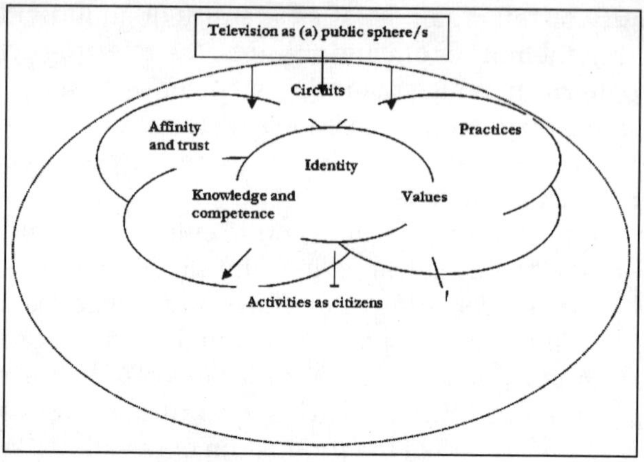

Fig. Circuits of Civic Cultures

Television as a 'Platform' for Plural Public Spheres

The final emphasis I wish to add to Dahlgren's model stands close to the discussion of the position of the viewer, namely the possible plurality of spheres. This is often seen to emerge regarding forms of and communicative modes offered by popular culture, for example tabloid television. Based on their work on talk shows, Livingstone and Lunt (1994) argue that the elitist, state-governed broadcasting as well as the market-driven commercial broadcasting both disregard the participatory aspect but, in concrete terms, 'a]s more ordinary people participate in making television programmes such as talk shows as well as receiving them, this gives new force to the concept of the active viewer'.

In their view, this creates alternative spheres next to, or opposed to, the traditional public sphere. Underlying here is the fundamental debate regarding the framework of the public sphere that Livingstone & Lunt have articulated as follows: Habermas proposes for the institutional development of a critical voice within the life-world so as to produce a consensual and reasoned challenge to established power, whereas proponenets of the oppositional public sphere, often drawing from the cultural studies tradition, view institution as a context for diverse voices with different power bases. My reading of Dahlgren's theorization makes it possible to consider that in this framework, amongst other components of a public sphere, television provides a particular platform for multiple spheres. What positions they take and in which kind of sense-making, consuming or will-formation processes they engage is not only their choice as individuals forming active audiences.

What is left from the original Habermasian public sphere concept, if I recognize the fluidity, plurality and heterogeneity as well as the limited potential of television as a public sphere? Again, by emphasizing the possibility of multiple, plural and changing public spheres, I do not mean to, for example, transform the public sphere framework into some 'poststructurally inflected version of hegemony theory' – as Glynn (2000) notes when critically assessing Livingstone & Lunt's (1994) revisionist terms of 'proletarian' and 'oppositional' public spheres in relation to tabloid television. What I have sought to stress is that change is not ahistorical, that nation-state bound, public service oriented broadcasting is still alive in Europe, but that the current social context also has elements of fluidity and plurality.

The possible existence of multiple public spheres is not a novelty but has been given importance when theorizing politics in relation to formation of 'new communities' and the new media. They form a plural and decentered public sphere constituted not by total identification as in the traditional community – for example, that of a nation-state - but rather by conflict, combining notions of interest and identity. It is characterized by the overlapping of an array of institutions and practical formations, including informal movements and associations where solidarities are formed, as well as formal governmental and legal institutions. Next to techno-utopian visions, it has namely been questioned if people's ability to create their own Net-based communities actually allows them to ignore other communities and 'bypass any contact with the balance of society'.

7

Television as a Cultural System

To examine how the above perspective on culture could be applied in intercultural communication research, let us take a closer look at how the three empirical dimensions — a worldview (Weltanschauung), a value system, and a system of symbolic representation — could be distinguished in intercultural television. Our starting point is the essay Religion as a Cultural System by Clifford Geertz (1973). Since James Carey (1975) pointed out the usefulness of Geertz's The Interpretation of Cultures (1973), this interdisciplinary book is also becoming a classic work for communication scholars. It may be seen as one of the cornerstone books contributing to the interpretive turn in communication research. The essay Religion as a Cultural System has a relation to television and could be adapted to Television as a Cultural System.

According to Geertz, religion is:

- A system of symbols which acts to...
- Establish powerful, pervasive, and long-lasting moods and motivations in men by...
- Formulating conceptions of a general order of existence and...
- Clothing these conceptions with such an aura of factuality that...
- The moods and motivations seem uniquely realistic.

Ernst Cassirer's adoption of the biologist Johannes von Uexküll's functional circle for the biological world showed us the centrality of a symbolic system in the human world. The answers in human beings toward outward stimuli are not direct, as in the animal world, but 'delayed' by the symbolic system. This symbolic system is therefore the key to understanding the differences in human culture. Differences in human culture arise out of differences in symbolic systems, which are made observable at the surface in a system of symbolic representation. Each symbolic system is culture bound, and television, like every other medium, is not only communicating through a symbolic system, but made up entirely by it. Consequently, the 'world of television' could be seen as a system of symbols, a point which has been made by many others. However, Geertz, for analytical purposes, adds an interesting double aspect to religious symbols. This double aspect is also of value for the study of symbols in the 'world of television'. First, they are a model of 'reality' (the 'real world'). In other words, they are an expression of a worldview. Second, they are a model for 'reality', which means that they provide some kind of blueprint for acting in accordance with that worldview.

However, the system of symbols not only refers to the media content, but also tends to identify television as a process in society, in which all the actions surrounding television are included in a system of symbolic representation.

Geertz's second point is a linear one as it draws on establishing motivations. Motives are neither acts (that is, intentional behaviour) nor feelings, but liabilities to perform particular classes of acting, or to have particular classes of feeling. The difference between moods and motivations is that where the latter are, so to speak, vectorial qualities, the former are merely scalar. There is an interpretation difference between moods and motivations. Motives are interpreted in s of their consummations, whereas moods are interpreted in s of their sources, without direction. Moods are shared, and in this sense communication can be perceived as a ritual act. Moods refer to the subjective interpretations of the medium and its content.

At the same time the symbols in the 'world of television' formulate ideas about a general order of existence. 'Man depends upon symbols and symbol systems with a dependence so great as to be decisive for his creatural viability and, as a result, his sensitivity to even the remotest indication that they may prove unable to cope with one or another aspect of experience raises within him the gravest sort of anxiety'.

According to Geertz there are at least three points where chaos threatens to break out in human beings:

- *At the limits of his/her analytic capacities*:

  What is important here is that human beings are always trying to explain phenomena, give them a place within their own worldview, rather than leave them unclarified or unclassified. The symbolic world of television and the surrounding system of symbolic representation could be seen as providing this possibility.

- *At the limits of his/her powers of endurance*:

  Like religion, television helps in dealing with the problem of suffering. Television confirms the existence of suffering, though, distinct from religion, it does so by making it observable and common, which also means shared, and not by making it understandable, explainable, comprehensive or one's destiny.

- *At the limits of his/her moral insight*:

  With regard to good and bad, justice and injustice, the symbols from the world of television build upon an accepted value system, which also provides a framework for dealing with chaos.

The fundamental question here is 'How is it that the denial comes to be believed?' In the study of religion this seems to be a little more difficult than in the study of television. Television has a higher level of reality than religion and therefore it is easier to believe. The 'world of television' seems to be uncoded, makes more sense and refers more to common sense or, in Habermas's s, refers to the common definitions of reality.

Geertz is unravelling the religious process and asking some fundamental questions about religion which could be important for the study of media in general, and the study of television in particular. The central idea is that television could occupy the same position as religion, myth and ritual in society. Consequently, the assumption is that it could be studied in similar ways. At the heart of these comparisons is the structural-functional position of television in a particular society or community, with culture being one of the most important directing factors. In this context, comparable with 'Art as a Cultural System', it is the talk about the phenomenon that needs to be given more attention and not only the talk about the recognisably aesthetic, which in television means 'talk about interpreting media content', but also talk which relates the phenomenon to wider social-cultural issues in society. An interpretive study of television must trace its significance in society, not in interpreted presented abstract:

The feeling a people has for life appears, of course, in a great many other places than in their art. It appears in their religion, their morality, their science, their commerce, their technology, their politics, their amusements, their law, even in the way they organize their everyday practical existence. The talk about art that is not merely technical or a spiritualization of the technical — that is, most of it — is largely directed to placing it within the context of these other expressions of human purpose and the pattern of experience they collectively sustain.

Therefore, when changes in the structural-functional position of television are identified, the analysis of the generation and transformation of ideologies begins. In his essay Religion as a cultural system Geertz formulates an important reason for studying the anthropological literature for this purpose: The anthropological study of religion is therefore a two-stage operation: first, an analysis of the religion proper, and, second, the relating of these symbols to social-structural and psychological processes.

The problem with contemporary social anthropological work in religion (including myths and ritual), according to Geertz, is that it is neglecting the first ('religion pure') and is concerned only with the second stage ('applied religion). It seems to us that in communication it is exactly the other way around. There are a lot of text-centred studies concerned with the construction and internal logic of the world of television, but only a few applying that knowledge at an empirical level to the relation between 'the world of television' and the social-cultural structures in society. Therefore, anthropology can be useful in this context. The mass media and television are cultural phenomena, rather than phenomena in a context of culture. Therefore, a perspective that takes culture to be an individual or collective reference framework, rather than a context in which media are operating, needs to be adopted.

INTERCULTURAL AND LOCAL TELEVISION

From the above, it follows that there is an urgent need for integrating fields and objects of study and more generally for a convergence approach. Following a bottom-up perspective in the study of intercultural television, an integration with local television and the different local uses of video seems obvious. The starting-point is a community at the micro-level. Community studies and community development have a long history. As early as 1955, Hillery had collected 90 definitions of the concept of 'community'.

A community here is based on geographical and ethnic aspects and does not necessarily have to be a 'community of interest', though it does refer to a social-cultural organisation, with specific community-bound processes, which one could include in the concept of structure. In this respect it has much in common with Tönnies's 'Gemeinschaft' or Redfield's 'folk society'. Recently the concept of 'interpretive community' has been introduced by Lindlof (1988) and others in the mass communication context. In our opinion, the concept of interpretive community is not just another way to reconceptualise the media audience. Nobody thinks of himself/herself as being part of an audience, there is no sense of belonging, but each one does think of him/herself as being a member of a community. The notion of interpretive community is therefore a reconceptualisation of community rather than of audience.

Community identity alongside and overlapping a cultural identity not only structures and makes sense of the media reception moment, but tends to identify the media and television in particular as cultural systems, which in essence are processes in a particular society or community. Consequently, the fundamental question is one of its dynamic position in that particular community, a position which is intrinsically linked to power and culture, both internal and external.

To assume that local communities are representative of a whole society and/or culture is regarded as naive nowadays. A microcosmic perspective, 'a natural laboratory', 'a closed universe' (Mead), 'a primitive isolate' (Pouwer) or, as Geertz has put it, 'the Jonesville-is-the-USA' or 'Easter Island-is-a-testing case' perspectives are either after-generalisations or 'hineininterpretationen'. If there is generalisation, it should be within cases and not across cases. The problem that arises lies, as Shankman points out, in how to accumulate knowledge. If generalisation and comparison across cases are viewed as destructive or at least not relevant, how is progress possible? The answer is that theory-building is not a goal in applied communication research but, like all participatory research, a problem-solving approach. Therefore we advocate something between a substantive theory and a 'nook-and-cranny' anthropology, an interpretive anthropology grounded in particular cases. We do agree with Kuper, however, that 'the business of making sense of ethnography requires theoretical guidance', which means that there is a need for an interpretation or theoretical framework for analytical purposes. Such a framework is necessary not only to structure and guide data, but also to make one's own position as researcher explicit. So, it has little to do with grand theory, but everything with avoiding guesswork.

Ten years ago, the proposal that cultural phenomena should be treated as significative systems posing expositive questions was a much more alarming one for social scientists — allergic, as they tend to be, to anything literary or inexact — than it is now. In part, it is a result of the growing recognition that the established approach to treating such phenomena, laws-and-causes social physics was not producing the triumphs of prediction, control, and testability that had for so long been promised in its name. And in part, it is a result of intellectual deprovincialization. The broader currents of modern thought have finally begun to impinge upon what has been, and in some quarters still is, a snug and insular enterprise.

[The study of society] is certainly becoming more pluralistic. Though those with what they take to be one big idea are still among us, calls for 'a general theory' of just about anything social sound increasingly hollow, and claims to have one megalomanic. Whether this is because it is too soon to hope for unified science or too late to believe in it is, I suppose, debatable. But it has never seemed further away, harder to imagine, or less certainly desirable than it does right now. The Sociology is not About to Begin, as Talcott Parsons once half-facetiously announced. It is scattering into frameworks.

Traditional positivist-functionalist approaches still implicitly start from the assumptions that all knowledge is based on an observable reality and that social phenomena can be studied on the basis of methodologies and techniques adopted from the natural sciences. However, as Giddens (1979) eloquently points out, the social sciences differ from the natural sciences in at least four respects: with regard to the study domain the social sciences, unlike the natural sciences, are in a subject-subject relationship; they deal with a pre-interpreted world in which the meanings developed by the active subjects form part of the production of that world; the construction of a theory of society therefore necessitates a double hermeneutics; and lastly the logical status of generalisations in the social sciences differs from natural scientific generalisations. Therefore in our opinion the social sciences are hermeneutistic and nomological in nature and need to be approached from a critical perspective.

We advocate the relative autonomy of a cultural analysis. In general, one can distinguish between two basic types of cultural critique. The first is of a philosophical nature, posing as an epistemological critique of analytical reason, of the Enlightenment faith in pure reason and in the social progress that rationality is supposed to engender. This type of critique attempts to demystify power and ideology. The second approach uses more empirical and therefore more conventional social science techniques to analyse social institutions, cultural forms, and the modes of discourse in social life.

Research Perspective

What we have in mind is a text that takes as its subject not a concentrated group of people in a community affected in one way or another by politico-economic forces, but the 'system' itself — the political and economic processes, encompassing different space and time constraints. This kind of analysis, which builds on the ideas of Giddens (1984) and Marcus and Fisher (1986), should also involve the relative power-linked articulation and conflict over ideologies, worldviews, moral codes, and the locally bounded conditions of knowledge and competence. Although all social research presumes a hermeneutic moment, often it remains latent because researcher and research inhabit a common cultural milieu. Moreover, it is in the study of the unintended consequences of action and the creation of meaning that some of the most distinctive tasks of the social sciences in general and communication studies in particular are to be found. At least two types of unintended influences can be distinguished: the unconscious ones, and influences conditioned by the context in which the different forms of social action take place.

In this context it seems almost unnecessary to say that we advocate a holistic approach. However, the media should not be central in such a holistic approach. According to Arbind Sinha, a holistic approach in communication research was introduced while evaluating SITE in India. But in anthropology this approach is as old as the discipline itself. The media must be studied in a geographical and socio-cultural defined area to accomplish a holistic approach. It is the socio-cultural framework that defines the ideological/mythical interpretations. The position of television in society is deined not only by other media such as radio or print (media paradigm) but also by other forces, powers and institutions such as economics, politics and even religion, art, science and so forth (institutional paradigm). There is no way one can grasp the whole field, but what is important here is that there should be an awareness that the media cannot be studied in a vacuum or only in relation to other media. 'Holistic' means that an element in a structure must be studied as an element in that structure, with direct integration of function.

The media in general and television and video in particular should not be central in such an approach and this goes one step further than arguing that the media should not be studied in isolation. What should be studied is culture. Culture must occupy the central place in research and should not be seen only as context. Focusing on culture then means selecting democratic interaction research topics (for instance in a funnel approach). Television is seen as an intervening factor in culture and we take culture to be responsible for the differences in the structural-functional positions of television in different societies. In this way culture is providing a reference framework for the interpretation of television as a process at a community basis.

To illustrate such a microscopic perspective, which integrates the different kinds of television from the bottom up, let us take a closer look at the Pacific Islands region. Little is known about the use of television and video in the Pacific Islands. The few things known about video indicate that all the islands have video, that it seems to be very popular, that movie-theatres closed down because of video, and that the governments are having great difficulty controlling the video industry. Broadcast television, on the other hand, at the moment is at different stages of development. Many islands are currently considering the introduction of television. Niue, the Cook Islands and Nauru recently did introduce a service provided by Television New Zealand.

Watching television in the Pacific is not an individual activity nor is television a family medium, but is viewed on a community basis. For instance, in most villages in Fiji there are one or two video-linked TV sets. The village leaders will regulate the viewing times so that they do not interfere with important village events. The same is true of the Federated States of Micronesia and, we assume, other places in the Pacific. This means that television could be studied as a community medium, which opens the way for the different perspectives on integrating the concepts of culture, function, structure, meaning, ideology and power — away from individual and psychological function and intellectual structure toward social function and social structure, away from programme content as text and toward communities and cultures as texts.

Research Design and Methodology

Without disqualifying and underestimating the significance of other research contributions we therefore advocate a research design that starts from a more dialectic and multi-centred perception of power factors in the context of communication for social change. In other words, we advocate a framework that also takes counter-power or empowerment from a bottom-up or grass-roots perspective into account. In general, one can distinguish among three problem areas in power relationships: the mutual dependency between the macro-level of the society or a given system and the micro-level of the social actions involved; the position and the autonomy of the subject; and the relationship of domination, dependency and subordination of power and interest contrapositions. The main actors of this new perspective are social movements and communities with a concern for multiple public issues. Therefore, instead of one central, objective, and mainly economic 'conflict', several segmented, subjective and 'postmaterialistic' issues can be identified.

Instead of one central collective actor (the proletariat or exploited class), several different, sometimes opposing collective actors can be identified.

This research project must centre on two problem areas. First, it must deine what actors or interest groups and what factors or structural constraints exercise influence from above. These influences can transform, reinforce, or weaken each other. What is required is a much more precise analysis of influence patterns that function from the top down by means of power in the broad sense. The second problem area is the grass-roots reaction to this influence. Research must be focused on the rational objectives of target groups and social movements. The difference from traditional anthropological research should be that the choice of the symbolic order for the research is deined by key concepts such as reproduction and labour. It is no more or less chance differences in rational objectives that are interesting, but the systematic tendencies and the thereby generalisable differences. This implies that the choice of the place and the context of research cannot be random but must be based on macro-structural insights.

Participatory Strategies

This policy and research model, which builds upon the multiplicity paradigm, starts from more dialectic mobilisation and conscientisation strategies. Social resistance and conscientisation are developed from the bottom up via interpersonal communication or media channels. This theme immediately brings to mind the conscientisation method developed by Freire (1970) in Brazil. Because Freire himself clearly states that the 'Third World' is also found in industrialised societies (what is involved here is not a geographical but a relational and structural categorisation), the 'exemplarisch lernen' method (imitative learning) of Oskar Negt (1971), which originated in Western workers' situations, can be mentioned as an important addition.

Freire's conscientisation method is oriented to bringing individuals to critical reflection about their own living conditions whereby they break through the critique-free adaption — the 'culture of silence' — and actively participate in a historical process. The right to participation and emancipation regarding the social, cultural, and historical reality is a fundamental right for everyone. The basis of the conscientisation method forms the philosophical and sociocritical notion that individuals must be able to achieve their essential goal — being a subject — and that the social structures, social relationships, and interpersonal relationships that interfere with it must be changed. The absence of any form of a guideline for political organisation of those who are conscienticised is partially resolved by Negt. In his method he tries to bring the marginal people in the society to see their individual need as a collective need that can only be satisfied in the context of organised groups.

In the literature this approach is often described as 'participatory action' or 'participatory research'. An appropriate definition used by the Philippine

Partnership for the Development of Human Resources in Rural Areas (1986:1) reads as follows: Participatory research is an alternative social research approach in the context of development. It is alternative, because although business, government and the academic also undertake research with development in the end view, little thought and effort go as to how the research project can be used for the benefit of those researched. The central element of Participatory Research is participation... It is an active process whereby the expected beneficiaries of research are the main actors in the entire research process, with the researcher playing a facilitator's role.

Participatory research calls for upward, transactive, open and radical forms of planning that encompass both grass-roots collective actions (that is planning in the small) and large-scale processes (that is planning in the large). This kind of planning and research is centrally conceived with human growth, learning processes through mobilisation, and the basic aim is to involve the people under study cooperatively in the planning and research process, with the planner or researcher acting as a facilitator and participant.

A 'NEW' PERSPECTIVE ON CULTURE

There are three basic assumptions that can be made about culture:

- Where people live together, there is culture (there are no societies without culture).
- Different societies have different cultures.
- Individual human beings in their personal development are highly influenced by culture.

However, this 'influence' should not be interpreted in a deinistic way. In other words, culture is a complex phenomenon that can be interpreted in a narrow or broad sense. We reject the viewpoint that conceives culture as only a partial phenomenon of the social reality and thus considers cultural sociology a subdiscipline of the social sciences. We believe that cultural sociology derives its special character not so much from a specific object field, but from a specific perspective on society.

In this context, culture can be defined as a social setting in which a certain reference framework has taken concrete form, or has been 'institutionalised'. This framework orients and structures the interaction and communication of people within this historical context. Therefore, the classic distinction between structure and culture as an empirical duality becomes meaningless. All structures are cultural products and all culture gives structure. This intrinsic bond with a society in which actions are full of value makes all social facts cultural goods. For social facts like institutions, behavioural patterns, norm systems, and societal models are construed and cultivated in the light of certain values, preferences or options that have developed in a society in response to certain common needs or problems. Those who are familiar with newer culturalistic approaches may observe that we build heavily upon the 'old' views of people such as Peter Berger and Thomas Luckmann (1967), Marshall Sahlins (1976), Raymond Williams (1980, 1981), and — of course — Clifford Geertz. The cultural process is linked to power in conscious and unconscious ways; it is sporadic and ubiquitous and transcends national and cultural boundaries:

Culture is the deposit of knowledge, experiences, beliefs, values, attitudes, meanings, hierarchies, religion, timing, roles, spatial relations, concepts of the universe, and material objects and possessions through individual and group striving. It concerns decisions not only about good and evil and so on, but also the way we eat, live, or dress. Therefore, various cultures also manifest different identities.

Three empirical dimensions can be distinguished in such reference frameworks: a worldview (Welt„anschauung), a value system, and a system of symbolic representation. The concept of culture implies material and immaterial aspects of a certain way of life, passed on and corroborated via socialisation processes to the members of that society. Ideological interpellations are made all the time, everywhere and by everybody. However, even though ideological interpellations occur everywhere, they tend to cluster at nodal points in the ongoing social process, which one could call ideological institutions or apparatuses.

Ideological institutions or apparatuses (for example school, media, church, the neighbourhood) are forms of behaviour that are crystalised, on the basis of social acceptance, into more-or-less standardised self-evident routines which can work as negative-repressing and as positive-liberating. They exist in strategies of relations of forces supporting, and supported by, types of knowledge which are discursive and non-discursive. They form clusters of institutions that have an impact on and influence each other and are distinct from others by their own identity.

The cultural identity refers to two complementary phenomena: an inward sense of association or identification with a specific culture or subculture; and an outward tendency within a specific culture to share a sense of what it has in common with other cultures and of what distinguishes it from other cultures. Within a culture we stress similarity rather than difference; to maintain the culture as separate from others, however, we stress differences rather than similarities. Like all social processes, these processes are not purely rational or pre-planned events. Thus culture must be seen as the unintended result of an interweaving of the behaviour of a group of people who interrelate and interact with each other.

In sum, culture is a phenomenon whose content differs from community to community, generally because the living conditions of these societies differ. Consequently, each culture has to be analysed on the basis of its own 'logical' structure. Each culture operates out of its own logic. In each culture one must therefore focus on the so-called archetypes rather than on the formal and often officially propagated manifestations of a culture. Consequently, there is also another view needed on what cross-cultural actually stands for in intercultural communication research. Building upon semiotic approaches to cross-cultural research a strict distinction needs to be made between culture in media content, and culture in everyday life.

The two cannot be understood in a direct causal relation and cannot be analysed and compared at the same level, simply because they are not of the same nature. Furthermore, in most cases contacts in cross-cultural communication are linear. This is especially the case for American television programmes in Third World countries. This means that often there is no feedback, no interaction, no two-way communication and no cultural exchange. Therefore, cross-cultural and intercultural communication are successful only when the above logical foundations are understood and accepted as 'equal' by the people concerned.

TELEVISION PORTRAYALS OF THE CRIMINAL JUSTICE SYSTEM

In general, crime and law enforcement television programs have been extremely popular, with consistently high ratings over time. More than a quarter of all prime time shows from the 1960s to the 1990s have focused on themes of crime or criminal justice, which constitute the largest single subject matter on television today, across all types of programming. Obviously, the nature of "television reality" is a key issue for this investigation. Numerous diverse and complex factors influence the types of programs and their contents that are shown during prime time, e.g., societal norms, values, and laws; agencies such as the Federal Communications Commission, Congress, and advocacy groups; the organization and interests of the broadcasting industry itself; and viewer attitudes and values.

Messages transmitted by crime and law enforcement programs must be explored to determine a number of matters — e.g., whether they are plentiful and accurate sources of information on legal processes, rights, and procedures; whether the legitimacy of the legal system is affirmed; whether rights are supported or are portrayed as an obstacle to effective law enforcement; whether the courts and the police are portrayed as effective, honest, or repressive; or whether the system is portrayed as corrupt, biased, or worthy of support. Some research has been done along these lines, providing broad (and, at times, conflicting) analyses of the content of crime and law enforcement programs. Drawing on Carlson's review of the literature, we see that these studies have typically covered five primary interrelated areas: knowledge of and information on the system, compliance, rights, police images, and violence and victimization. Each of these substantive areas can provide guidance in developing a comprehensive research program focusing on television imagery and public perceptions of the criminal justice system.

INFORMATION ON THE SYSTEM

As we have discussed, there are arguments to the effect that television, even when entertaining, is a powerful educational medium. For example, as Gunther notes, "there must be millions of people who have learned, simply by watching crime dramas in the past few years, that they have the right to remain silent when arrested." The American Civil Liberties Union has likewise said that television crime and law enforcement programs provide an "excellent legal education for the public".

However, on the other hand, many analysts find television programs to be poor sources of accurate knowledge regarding even the formal operation of the legal system, due to omission and distortion of information and the higher priority of dramatic necessity. For example, most police crime shows depict a criminal investigation process ending with apprehension, arrest, or violent death of the suspect. The stories often include legal and illegal searches and seizures, and arrested suspects are often "read their rights." Postarrest procedures — e.g., arraignments, pretrial hearings, jury selection, bonding, plea bargaining, trials, sentencing, etc.— are rarely shown. Even on courtroom drama series, these procedures are rarely seen; neither are instances where cases are resolved before going to trial. In addition, in reality, criminal trials seldom end in courtroom confessions as they frequently do for dramatic effect on television.

The subject matter of programs provide information about the types of issues confronting like characters in "real life." This point is even more dramatically expressed in so-called "reality-based" shows, of which there has been an increasingly burgeoning number in recent years. In actuality, these shows are typically, to a large degree, "programmed and staged" (even if only to accommodate technical requirements). Yet, they are presented as "real," and are often produced in a format and style associated with documentary or news reporting, about which questions of social presentation and accuracy can also be posed. Despite some descriptions as "tabloid" television, these shows are evidently perceived in the manner in which they are presented and thus, although currently not dominant in prime time viewing slots, require serious consideration in terms of the "information" they present for public consumption and perception.

Of course, there are also some very compelling arguments that maintain that television information and messages are, in fact, not meant to facilitate accurate learning about the criminal justice system, but rather are intended to make it less understandable and clear, in support of social control. Crime and law enforcement programs show television characters, not rules or laws, as solving the conflicts; i.e., "television personifies the law.... By portraying the law in this way, television makes the legal system and the very foundation of the law less understandable". This argument corresponds to assertions that television supports the wider system and ties into arguments that the legal system functions better with a relatively ignorant citizenry. Briefly, television programs on crime and law enforcement are typically characterized by omissions or distortions on information about the legal system. (I might also mention here that programs focusing on prisons are almost nonexistent in the overall scheme of television programming production and content.) In any case, while there are scenes where suspects are read their Miranda rights and search warrants are served, some researchers maintain that little other accurate information is forthcoming from television portrayals of the formal operation of the legal system. As Surette argues, "in every category — crimes, criminals, crime fighters, the investigation of crime, arrests, case processing, and case dispositions -- the media present a world of crime and justice that is not found in reality." Of course, more longitudinal documentation of related television program content is clearly required in this regard, to determine these trends over time and to provide a basis for further theoretical and empirical analysis.

COMPLIANCE TO THE CRIMINAL JUSTICE SYSTEM

The most dominant message of prime time crime programs is that "crime does not pay" and that criminals are highly undesirable and unworthy characters. Most shows portray a lawful life as clearly superior to an unlawful one, and almost all television criminals are caught or thwarted. This type of message can make a powerful contribution to general social control and compliance. Carlson (1985) denotes three characteristics of crime and law enforcement programs that contribute to messages of compliance as the most appropriate behavior:

- The sharp distinction drawn between criminals and law enforcers,
- The often violent consequence of criminal activity, and
- The effectiveness of television law enforcement.

First, television criminals are often given a wide variety of "negative character traits" in addition to their criminality, and their motives are typically greed or base wickedness or cruelty. On the other hand, law enforcers are depicted as relatively moral. Although they may be independent and conflict with the bureaucracy, they are always clearly on the side of the law and conventional morality. Character depictions are highly important, especially in regard to broader social attitudes and relations. The character attributes convey messages about different "kinds" of persons and segments of the population. In terms of social learning, these messages help to define the function, usefulness, abilities, and other features of various groups of people. These characteristics need not be obvious. Indeed, "we are usually not aware of the relative shadings of each role because each is rationalized by the particular plot.... We are even less aware of the associations common to large numbers of characterizations that we do not perceive to be parts of a wider pattern".

Second, violent resolutions to criminal activities are overrepresented in television crime programs. A violent end makes clear who is winner and who is loser, who is superior and who is inferior. Third, while high levels of police and court effectiveness in television programs may be unrealistic, their portrayal as such arguably encourages certain perceptions of the system in operation. Police almost always get their way; crimes almost never go unsolved; murderers are almost always caught; and courts and juries almost never let crooks go free. In short, crime does not pay.

Compliance is clearly central to television portrayals of crime and law enforcement. However, it has been argued that, while television programs ostensibly support legal compliance, the real message is that people, not the law itself, insure compliance. An emphasis on personalities resolving disputes, rather than the law, obscures the very basis and rationalization of the legal system. Television apparently presents a distortion of reality in terms of both compliance and the nature of the criminal justice system.

DUE PROCESS AND RIGHTS

Television emphasizes repression of criminal conduct as the most important function of the legal system, in support of the social order. For example, in a typical scenario of police shows, the viewer sees the crime being committed and knows the identity of the criminal. Thus, when the police conduct an illegal search, it seems more "acceptable," or at least less questionable, than it might otherwise, since guilt has been established in the eyes of the viewer. In the same light, the viewer might feel that a television lawyer or judge goes too far in protecting a suspect's rights, with innocent people suffering as a consequence.

Most programs portray illegal searches as essential, revealing a vital piece of evidence; if police brutalize witnesses, it is to obtain a crucial lead for the capture of vicious criminals. The disparity between the televisually constructed reality of the criminal justice system and the actual reality means that the public receives an image of crime and justice that typically supports a policy of crime control mechanisms ahead of due process protections. Television portrayals of violations of constitutional rights are typically not identified as such, and even if so, are depicted as necessary for effective crime control and system maintenance. In this way, whether viewed as psychological manipulation or not, public perceptions and knowledge of the actual criminal justice system can be affected.

PORTRAYALS OF POLICE

Of course, television images of police can be expressed along a number of dimensions. Most television law enforcers are directly connected with the police, courts, or a government agency, and they are overwhelmingly white, male, and middle-aged, although it might be that over time the age category appears a bit more flexible, occasionally shifting between "youngish" and middle-aged white males. Furthermore, not surprisingly given dramatic "requirements," television unrealistically shows police work as primarily investigative, generally ignoring routine police work; it is depicted as glamorous, with a great deal of independence and access to tremendous resources, with an extremely high rate of success. Moreover, in general, television portrayals of police are generally positive, showing them as honest and law-abiding.

VICTIMIZATION

Murder is the single most common crime on television. The other most common television crimes are robbery, kidnapping, and aggravated assault. "Most of the crimes are vicious attacks by truly frightening calculating criminals on innocent victims.... Violence seems to permeate all human activity and acts of murder, rape, kidnapping, and especially terrorism are on the increase". Some analyses show females of all ages, young boys, nonwhites, and lower or upper class persons are the most likely victims of violence, with female victimization rapidly increasing. In any case, overall, television reality posits a great likelihood of being involved in violent activity and crime can occur almost anywhere. On the other hand, the television likelihood of being a victim of violent crime is greater if one is young or elderly, white, lower class, and female.

However, the probability of television victimization is relatively high across all groups of the population. This type of portrayal might have a powerful influence on the viewing public's perceptions of reality in regard to the criminal justice system. In particular, it seems highly supportive of the status quo and conventional views of "proper" behavior and morality, necessitating extreme levels of crime control and latitude for law enforcers, and maintenance of the legal system. This point again suggests that television programming, as a cultural object, is not only reflective, but instrumental in supporting the ideologies and beliefs of mainstream, or dominant, society.

8

Division of Labour in Television Production

Television's language is evolving at great speed, and the medium reflects upon itself to a considerable degree. This self-reflexivity ensures that even casual viewers can remain fluent in their understandings. Popular understanding of how TV is produced is necessary if viewers are to know the status of what they are being shown, not so that they can go out and make TV themselves. They need to identify what is live and what is recreation, what is fiction or fact. They have to distinguish between contemporary or archive footage. They need to distinguish between the different levels of factuality claimed by particular footage: literal truth or the statement of a particular point of view. They will come to a judgement about whether an interviewee is telling the truth or whether they have been 'exploited'. They need to make correct generic identifications and to measure a particular programme against generic models.

Modern TV audiences already have an understanding of TV production, at least in the developed world. TV's reflexivity is extraordinary, ranging from the bloopers of the stars to the instant recycling and commentary upon the latest piece of startling news footage. Mix in the commentaries on TV in related media, radio DJs chatting about last night's 'must see' show or newspaper gossip and revelation, and you have a medium which is exceptional in the way that its production processes are open to the world. Virtually everyone knows its jargon and production secrets; could we say the same about plastics, waste disposal or even education? Indeed, such is the self-reflexivity of TV and the extent of reflection on TV in other media that many have rushed to see it as the 'post-modern' medium par excellence.

This is to miss the point. The only reason why television is reflected upon, why its production processes have become common knowledge, is the universality of the medium which has gained the currency of everyday life itself. To assure significance in the developed world today, any new phenomenon has to touch and be touched by television. That is, it must be touched by the dominant form that television takes in modern society: the form of broadcast television programmes supplied into people's homes. Other uses exist, from the seemingly automatic forms of surveillance cameras to the highly wrought forms of video art, and all have their own organisations of production. Yet so powerful is the dominant form that the term 'television' is virtually synonymous with its broadcast form, especially if we see 'broadcast' as including all scheduled, linear services, including cable and satellite services.

Programme product circulates relatively freely within this constellation of channels, and production processes are similar. US cable shows like HBO's Six Feet Under appear on free-to-air broadcast channels in other countries (eg Channel 4 in the UK) with advertising breaks. In commodity terms they are no different from other programmes on those channels. To produce a series for HBO rather than the networks in the USA does not involve a fundamentally different production process, except in the important respect that there is a greater degree of editorial freedom, which is partly the result of a degree of flexibility in terms of the size of series runs, budget level and so on. Hence this article will use the term 'broadcaster' to cover a wide range of organisations interested in obtaining TV programmes for broadcasting over networks to a substantial domestic customer base, be that through free-to-air distribution or through subscription.

Broadcast television defined as scheduled linear services for a domestic audience is a universal given in the first world. Self-reflexivity is the by-product of this universality. Self-reflexivity in television is the equivalent of word play and jokes in language, a constant attempt by speakers to assert their control and creativity within the vast and universal system of language. But there is a crucial difference here. In language, all listeners can become speakers, even if the power relations in a particular circumstance might radically restrict what they can say. But, up to now at least, those who receive television have little or no possibility of becoming producers of television utterances. So the general understanding of television remains partial. It does not take viewers very far in understanding the routinised aspects of television; the cultures that produce its particular forms; or its attempts to regulate supply and demand. Television's backstage is less glamorous than it might seem from television's own representations of itself.

ROUTINE

Television production has become routinised because television watching has become such a central part of everyday life. Television can mount special events, but the overwhelming quantity of television consists of series and established formats, and indeed that is what all but the most old-fashioned of commentators expect of it. The routines provide security, both for the industry and for its audiences. As more channels appear, the routinised aspects become more rather than less marked. The early years of television in the middle of the twentieth century, sometimes referred to nostalgically as a 'Golden Age,' saw what appears now to be a remarkable number of single plays or films (in series with titles like the Philco Television Playhouse, Armchair Theatre, Wednesday Play, Comedy Playhouse). This early period lasted into the Seventies in Europe, with the production of short-lived but much loved series (Fawlty Towers for instance). Who now would waste such ideas on a single outing, except, of course, those who are producing high-profile events, high-end drama or 'TV movies'? It is no longer possible for single episodes or short runs of more routine TV production to win a place in a marketplace already crowded with excellent and/or loved material. Television is a victim of its own ubiquity. Its principal means of giving currency to its own new productions is to establish them as a fragment of the everyday realities of their audiences. So television becomes routinised even at the point of creativity.

Routinisation has its advantages for producers. It brings considerable certainties and efficiencies to the production process, which is fraught with risk and creative wastage. It means that each production is made within a framework of common understandings. Each genre has its normal ways of doing things: 'gameshow' means a studio-made event with an audience, celebrity presenter and contestants. It means a ritualistic process with its own catch-phrases. It also means a long series and a relatively brash visual and aural style. Imagine trying to explain all of that to a broadcaster before you can say what makes your format different; imagine having to define the genre for the technicians, the designer, the presenter, the gag writers. Specific formats are difficult enough to work out and define in any case, however simple they may look on screen.

Some degree of standardisation is necessary in any industry but there are few global standards in production for broadcast television. Cultural specificity and the need to adapt successful formats for local conditions have long been crucial aspects of the industry. The makers of the successful US gameshow Wheel of Fortune were able to roll it out into many different countries. But they did not only try to sell the original shows they made for US television. They also licensed the format to broadcasters and/or production companies in Germany and France, etc to make local versions of the format, with local presenters and contestants, local gags and prizes appropriate to the market. They also allowed local production companies to alter the running time and even budget level and production values to suit national broadcasting conditions.

Standardisation

Considerable standardisation does exist in broadcast television, but it is different in different nations and regions. A crucial part of the skill of a producer is to know those standards, and a skilled producer can 'read' the probable level of the production budget just by looking at a TV programme produced in their own market. This informal standardisation is often reflected in the 'guide prices' that are issued by broadcasters when they put a slot or format out to tender in the production community. The price is only a starting point for negotiation on the chosen project, but to tell a producer that a drama will cost 'x' per episode rather than '2x' is to send powerful messages about the expected nature of the product. Running time is another significant method of standardisation.

Broadcast TV has standardised into slots of 30 minute units, with very rare exceptions. A producer or director claiming that their project has a "natural duration" of 11 minutes per episode is undertaking a form of commercial suicide, unless they are addressing a broadcaster with a particular need. Such needs can include material for a magazine format designed to fit a 30 or 60 minute slot, a larger entity like the 'open plan' children's TV sequences, or to fit the gaps left in the slot structure by a feature film.

The pattern of slots based on 30 minutes is practically universal in scheduled broadcast TV: it provides the basic building block of the schedule. But the precise duration of the programmes that fit those slots varies according to the practices of different broadcasting markets. The US network primetime slot is around 44 minutes; the UK commercial network slot is around 51 minutes; for the BBC it is more like 56 minutes. The difference is the permitted level of commercials (and in the case of the BBC, the lack of commercials). Hence drama series made for the BBC have to be cut for US transmission, except on channels like PBS which carry no commercials.

In the UK, it sometimes creates space for short items; sometimes a schedule of two US products back-to-back; sometimes a plethora of trailers for forthcoming shows. Co-productions are sometimes hampered by this lack of standardisation: the BBC wildlife series The Blue Planet and the Life of Mammals were made in co-production with Discovery Channel. In order to make the episodes up to an hour slot for BBC transmission, an extra segment was added to the end of each show, detailing the intricate and ground-breaking production methods behind some of the spectacular sequences. A new level of reflexivity was added to the programmes.

Production standardisation is a therefore a feature of each production context rather than a universal. Within each production context, it underpins the organisation of production, enabling workers within the production community to move easily between projects for the wide range of outlets within the national broadcasting market.

It enables crews to come to a common understanding quickly about the scope and ambition of the project they are working on; it enables broadcasters and producers to have meaningful conversations about creative ideas. Finally, it enables production work to become more collectivised, leading to developments such as team scriptwriting. To become a key production worker within a particular market, it is essential to acquire a sense of its underlying culture of standardisation.

DIVISION OF LABOUR

Television production has a factory nature, and as in many modern factories, most of the workers are hired on a casualised basis, to complete a job or project. When compared to the feature film industry, much work in television is standardised, repetitive and anonymous at most levels. But it still depends upon the creativity of the people who work within in. This creative process is organised into different levels. Writing an episode of a soap is different from devising the series as a whole. But the people who devised the soap, who created it originally, came up with a format, a blueprint, and not hundreds of finished episodes. They will have written sample scripts, and will usually be heavily involved in the scripts for individual episodes for the early years. But what the devisers have created is a virtuality, a potential imagining of a programme which has to be shared if it is to come into being. This more collective basis of television production is difficult to reconcile with traditional views of authorship, even as applied to feature filmmaking. It depends upon a known division of labour, a hierarchy of creative responsibilities, and a sharing of creative activity. These are the result of product standardisation.

ORGANISATION OF PRODUCTION

It is convenient to divide TV production into five phases: finance, pre-production, shooting, post-production and marketing. From the point of view of any one programme, they appear to be a natural sequence: first you find the money, then you plan and shoot, then you edit, and then you distribute the finished product to an audience. However, each stage makes assumptions about the others. Changes at one point have profound consequences at another. For instance, in 1994 the BBC made the marketing decision that it should show more trailers as on-screen promotion of future programmes. Every programme, regardless of length, had to be shortened by one minute. Not much of a problem for 60 minute drama, but more of a problem for my series called French Cooking in Ten Minutes, which lost ten per cent of each programme a week before it went into the studio.

Production in each broadcast market (normally a national broadcast market) has a particular national character resulting from both its internal evolution and the impact of foreign practices upon it. Production organisations generally take three forms: the vertically integrated producing broadcaster, independent production and forms of co-production.

VERTICAL INTEGRATION

In the early days of live TV, especially in Europe, it was natural for broadcasting and the production of programming to be contained within the same organisations. Programmes were made and broadcast from the company's own studios. The only alternative was to produce work on film for broadcast later, and the costs of this were prohibitive until a substantial viewership had been created. Nevertheless, the live studio model of production continued to dominate many genres, and vertical integration continued to be a logical way of organising broadcasting. As the medium evolved, live studio production became a relatively specialised part of television production. Many broadcasters continued to produce a substantial amount of their own programmes, but as a result of the gradual increase in production values since the 1970s, almost all broadcasters have had to look beyond their own resources in a systematic way.

This has created a fundamental cultural shift within the organisation of vertically integrated companies, often referred to by the term used within the BBC, "producer choice". Essentially, the process is one of adaptation of a large vertically integrated company to the demands of full-cost budgeting, where everything is costed in terms of money rather than a mixture of money (paid to outside companies) and resource time (allocated by the vertically integrated broadcaster). The BBC came to this process relatively late, and because of its size and its cultural position as a primary definer of 'public service broadcasting' for many smaller markets (e.g. the Scandinavian nations), the process has been the subject of several academic studies.

The vertically integrated broadcaster has most of the equipment and studio space necessary to make its own productions. So the process of production was most conveniently planned by allocating resources by time: so many days in the studio; so many hours editing; a crew of a given size for a given number of days; a director for a specified number of days. This simplified internal resource allocation and gave producers a clear indication of what they had to work with. A separate cash budget was then drawn up for the money to be spent beyond the confines of the vertically integrated broadcaster: be it on talent (actors, writers, freelance workers) or materials (sets, lunches, archive material, car hire, costumes etc).

This model had many advantages; but knowing how much a programme cost was not one of them, as there was no charging structure for the internal resources. Over time, therefore, it becomes difficult for a vertically integrated broadcasting organisation to know how efficient it is compared to other production organisations. It becomes difficult to put a price on its own products; and impossible to compare costs with production companies working in a full-cost way, paying to buy or hire everything from other companies. The process of moving to a full-cost production model is difficult, as it implies that the resources allocated to producers would need to have a price put upon them, and a price that would enable the organisation to plan its own maintenance and renewal of those resources.

The BBC, under the Director General John Birt, adopted an aggressive version of this process which sought to price all activities internally, and after a time allowed staff producers to spend their budgets on resources chosen at will from within and beyond the BBC. This created an internal market, exposed to competition from external companies of very different natures, and therefore tended to erode the central back-up services (library, press clippings service, pronunciation unit etc), which create the economies of scale that are the characteristic of an efficient large organisation.

However flawed the implementation of this process, the principle was essential: in the long term, a vertically integrated broadcasting organisation has to be able to compare its own operations with those of other companies in the same field. This is a prerequisite of the necessary level of production standardisation within a particular production culture. As a direct result of the process, vertically integrated broadcasters were able to reduce their 'plant' and equipment costs and cut the number of fulltime employees. In doing this, they were participating in a widespread trend in modern industries towards the use of skilled freelance labour and the outsourcing of aspects of production to independent companies.

Independent Production

Independent production takes place when a broadcaster contracts with a separate company to provide it with programmes. It therefore takes many forms. An independent can simply be an outside company that provides ideas and talent but uses studios and other facilities owned by a vertically integrated broadcaster. Or it can be a totally separate company that proposes an idea, has it accepted, finds or even provides all or part of the finance, makes the programmes and simply delivers them to a broadcaster at the end of the process. Independent producers sometimes sell all the rights in their programmes to the broadcasters financing them (the model that still prevails in the UK at the time of writing); or sometimes keep specified rights to sell the programmes to other broadcasters in the same or in foreign markets. Independent producers sometimes own their own production resources; sometimes hire them from other specialist companies. Independents sometimes have their own talent under contract (actors, directors etc); sometimes hire them in on a project-by-project basis. There is only one thing that defines an independent producer: it is a company with no access of its own to a broadcast channel on which to show its products. An independent production company therefore has to make a contract with a broadcaster for this access, and therefore has to seek commissions for programmes from broadcasters. Even this definition is clouded in some markets, where broadcasters take a direct investment stake in independent production companies.

Co-production

Co-production involves the co-operation between companies on a specified project. Co-production usually takes place when one broadcaster cannot afford to produce something on its own, so has to seek financial or other contributions from other parties or may simply wish to share the costs and hence the risks associated with programme production. Co-productions often involve arrangements between programming producers or between producers and broadcasters, but may also involve banks or other investment organisations; commercial organisations, political or charitable organisations with a message to promote; or even the independent production company itself. In all cases the old maxim of 'nothing is for nothing' will operate. Where a broadcaster from another national market is involved, there may well be issues of cultural specificity or different running times or other issues of product standardisation to be resolved, as with the examples of The Blue Planet and The Life of Mammals above. Where outside finance is involved, the programmes will tend to be produced to maximise potential revenue both from the sale of the programmes and from spin-off products. Where organisations with a particular message are concerned, their motivation is at least clear and known from the outset.

PRODUCTION ORGANISATION

Production is organised differently in each of the major genres of TV. Situation comedies and series fiction (also known as drama) may appear to be similar in that they show fictional characters in a variety of narrativised incidents leading to a conclusion. However, their styles of production can be very different.

An established sitcom like Frasier has a standard pattern. There is a large writing team, consisting of a number of paired writers. For each episode, a pair will be delegated to write a first draft as part of the overall series planning. This will have planned the narrative arc of development for the characters during the series; the general themes of particular episodes; and any episodes with unusual set requirements (holiday locations etc). The show is heavily dependent on a small number of standard sets: Frasier's main apartment room being the chief one, along with the radio station, Café Nervosa, Niles's apartment and the other rooms of Frasier's apartment (kitchen, bedroom).

A normal episode will be allowed an additional special set, the requirements for which will have been planned in advance. Each Monday during production the pair delegated to the week's episode read their draft, and it is discussed by the writing team as a whole. Rewrites of specific scenes are allocated to other writers. On Tuesday these are brought together into a composite script which is again read through by the writers and given to a writer whose specific job is polishing: adding business and smoothing through the flow of gags. On Thursday the cast read through the script for the first time with the writers. If it works, the show is taped in the studio on Friday. If urgent rewrites are required, the taping is put off until Saturday. And on Monday they start over again, usually for 26 weeks a year. In the UK and

Europe, it is normal for sit-coms to be scripted by more than one person or even a team, and to follow something of this pattern.

However the degree of rewriting habitual in the US is unheard of in the UK. In the USA, team writing of drama series has also become a common practice, though it is rare elsewhere. In episodic series fiction like NYPD Blue or ER the activity of assembling a script is more complex because of the multiple story lines involved. European drama, prime-time soaps apart, is more often the work of a single writer and is planned in shorter seasons.

The shooting of sitcoms and drama differs markedly. Most sitcoms are made like Frasier, shot in studio sets with multiple cameras in front of an audience. TV drama never has an audience at its shoots, and – with the important exception of daytime soaps - moved out of the confines of the studio during the 1960s to adopt a much more filmic method of production centring on single camera shooting. The shooting of a 30 minute sitcom can be done in the space of about three hours. Thirty minutes of drama shot with a single camera would normally take six days, though tight finance, careful planning and routinisation (enabling the use of multiple cameras) can reduce this time substantially, as is the case with soaps. As production differs so markedly for different genres in different markets, any general statement about the nature of production has to be taken with caution. However, it is possible to distinguish between five broad stages.

STAGES OF PRODUCTION

Finance

TV production involves spending other people's money, and usually a lot of it. For example, the Australian first series of Big Brother cost approximately 16 million Australian dollars (approximately US$ nine million) - see the article by Jane Roscoe in this section. Unsurprisingly, those people like to have a say in what is produced, so the activity of financing a production often determines its most intimate details. In addition, there are few agencies outside the television or media business that are interested in investing in speculative TV production because of the nature of television finance. Purely speculative production (that is, producers spending money without a pre-arranged outlet) tends to take place for the cinema rather than for television. There are a number of reasons for this, besides the most obvious one that cinema is more glamorous.

The first reason is that television companies make expensively but buy cheaply. Broadcasters are prepared to pay more for programmes they commission than they are for ready-made material. There are two very different price scales for programmes depending on whether the broadcaster is involved in commissioning and/or making them themselves, or is buying them as completed programmes. A broadcaster might well commission a documentary costing £50,000 (approximately US$75,000) on the basis of little more than an idea and the director's track record. However, the same documentary offered for sale as a finished item by the same director would be purchased for, at best, £10,000 (approximately US$15,000). It all seems rather odd, as there would seem to be less risk involved in buying something that is already finished.

There are several reasons for this. The most important is that when a broadcaster pays for the production of a programme, they will own it (or a substantial share in it) as a commodity. The rights in the programme can then be sold over and over again to different broadcasters as well as to other outlets like video and DVD. However, programmes are rarely simply sold from one market to another for reasons of cultural specificity or the different nature of product standardisation in different markets. A broadcaster may, for example, use a particular known voice for documentary commentaries, so will wish to add this voice. They may well wish to rewrite the commentary to suit local conditions, both for range of references and for overall tone.

So the advantages of having a finished programme often have to be measured against the advantages of exclusivity (of being the first to show the completed programme) and creativity (of having 'an input', meaning telling the director to do certain things, and of being able to cater for cultural specificities). With increasing competition between channels, the issue of 'creating a unified look' or of branding has become increasingly important, permeating the look of individual programmes to a remarkable extent. Hence broadcasters tend to want to control such aspects of the production, and so put a premium on controlling the production process through investment in it. Purely speculative television production is therefore a difficult option for investors, outside areas such as children's programming where the brand is often the character or the toy which will become more valuable as a result of the TV programme. Where a producer is selling limited rights to a broadcaster (e.g. the right to show the programme three times over one year), or where broadcasters from different countries are involved, the producer may well end up making different versions of the same programme, complicating both shooting and post-production.

For both parties, the broadcaster's knowledge of scheduling, of how the programme will 'fit' into a slot in a sequence of programmes remains crucial, even for drama series. In addition, the television market place is full of programmes being offered for sale by broadcasters and program producers from other, usually international, markets, all of whom participate in the same financial model. In the case of smaller territories, the purchase of large amounts of foreign-produced material at low cost is what makes possible the provision of a broadcast service extending for hours and hours.

The second reason for the lack of speculative production in television is that the industry has a long history of vertical integration, of single companies undertaking both production and broadcasting. TV's early history as a live medium, where shooting and transmission were one activity, meant that it was natural for them to be contained in the same organisation. The one market that saw an early division of the two activities was the USA. The existence of four time zones made it desirable to record the nation's peak evening entertainment for delayed transmission to other time zones.

In the USA also regulation produced a situation in which local broadcasters were separate from the emerging national distribution networks and contracted to take all or part of their output from them, but tended to reschedule some of it for their own reasons. In the smaller territories of Europe, single national broadcasters emerged like the BBC in Britain or RAI in Italy. Subsequent developments in Britain tended to be variants on the broadcaster-as-producer model until the introduction of Channel 4 in 1982, which introduced the model of commissioning productions from independent producers. This model was widely adopted elsewhere during the subsequent decade.

THE COMPEXITIES OF US TELEVISION FINANCE

Production finance in the USA had organised itself along a radically different model. American television began with separation between networks and stations, providing a secondary domestic market. So producers could sell their productions once to a network and then over again to local television stations. This 'syndication' market opened up in the 1950s, enabling producers to sell filmed series, game shows and sitcoms to local stations to fill the endless hours of daytime and early evening TV which the network did not provide. Syndication provided material as well for those not infrequent spaces in the schedule that arose when stations affiliated to the network refused what the network offered, not least because they wanted to keep all that slot's advertising revenue for themselves.

The equivalent in European broadcasting emerged only with the arrival of multiple cable and satellite broadcasters in the late 1980s. Even then, European terrestrial broadcasters, usually the principal or only financiers of production, tended to hold on to the right to reshow programmes they had financed, rather than allowing producers to retain them. This was because they typically provided all of the production costs, whereas production for the American networks had long since been on a shared risk model: the network provided a proportion of the costs and the producer had to find the rest.

American TV production finance became as complex as feature film finance as long ago as the late 1950s. American networks would commission – and usually finance - pilots for series. On the basis of the pilots, successful programmes were commissioned as series. The broadcaster would pay a proportion of the costs to secure a number of showings of each episode over a delimited period of time. The producer would then raise the rest of the money, aiming for the lucrative income from successful shows with 100 or more episodes, which could then 'go into syndication'. Although such successes represented only a small fraction of all the shows broadcast on network television, the aggregate size of all the local markets in the US meant that a 'hit' show could generate hundreds of millions of dollars for its rights holder. Often, therefore, the non-network finance would come from distribution companies skilled in selling to the syndication market, and interested in the income from sales to TV stations outside the USA. These distributors are large enough to be able to spread their bets across a wide range of projects in the hopes that enough will go to syndication. This model of production worked well for long runs of filmed fiction and sitcoms, which is perhaps why these forms seem to dominate in the USA.

Other models emerged as well. Sometimes straight-forward sponsorship by advertisers would cover the difference between production costs and what the network would pay. Soap operas are so called because soap companies like Procter & Gamble originally produced the shows themselves. More recently, barter has been popular for cheaper studio-based programming such as talk shows, cookery or cinema review programmes. Network TV will pay the full production cost of such programmes, but in the syndication market the producer will split the income for the advertising breaks, selling half to national advertisers to cover production costs and generate a profit, and leave half for the local station to fill.

Such are the complexities of the US market. Other production markets have not yet arrived at such a system, but as the number of channels and the weight of independent productions both increase, such models will begin to become more typical. Both systems still involve one crucial phase for almost all successful productions of any size: the involvement of a substantial broadcaster from the beginning of the project. Nowadays this can be a subscription channel like HBO rather than a free-to-air advertising-supported network like NBC, CBS, ABC, Fox, WB or UPN, but it is crucial to have a contract with a major broadcasting company that will bring the production to viewers and seal its reputation.

Commissioning

The activity of obtaining a production contract from a broadcaster is known often called deal-making by producers and commissioning or ordering by broadcasters. The two are conflated within vertically integrated companies, and considerable tension exists when independent suppliers compete with broadcasters' own production units for commissions. In either case, the process has a number of steps, some of which can be skipped depending on the producer's confidence in their idea or their degree of contractual intimacy with the financiers.

What comes first are ideas and needs, but it is not necessarily the case that producers have ideas and broadcasters have needs. Producers can come up with ideas, but they also identify 'gaps in the market' or seek to exploit talent with which they 'have a relationship'. Broadcasters identify problems with their output and seek ideas for dealing with them, sometimes defining them very tightly. But they also cast around for the talent of the future and for the coming ideas, looking for the faces and genres that are coming into fashion. They are aware that they are in an ideas business and often generate their own.

Producers looking for finance for ideas will often research the output of channels to find the most suitable place for their ideas or scheduling weaknesses which their ideas can address. Sometimes producers will generate investment for projects before approaching broadcasters in order to make the proposal seem attractive or even viable. Audience viewing figures and more general market research are therefore habitually used at this point as a crucial tool in developing ideas. Ideas may sometimes be abstract at their moment of conception, but from there on in they have to assume very concrete forms.

Once a need encounters an idea, a deal can be negotiated. The deal is essentially a blueprint for production. It specifies the who, what, how, when and where as well as the how much.

Typical deals between separate organisations (producers, broadcasters, other financiers) will cover the following kinds of issues:

- The eventual look of the programmes produced, specifying cast and senior crew, length of shooting time, what technologies will be used, delivery dates, lengths of programmes etc.
- The ownership of rights in the programme, which are often of limited duration (e,g. terrestrial broadcast rights for five years), limited in territory (e.g. US and Canada only), and even limited in number (e.g. three runs over five years). Rights then revert to the production company or are passed on to other investors.
- Who has the right to control aspects of the content of the programmes, and at what point this right can be exercised. This includes issues such as the right to re-edit for particular purposes or the specification of different versions for different purposes to be delivered by the original producer (e.g. a specific European version to be longer than the US version, and to include material specifically addressed to the European market).
- The way that the finance is structured: who will pay what when, and what happens if they don't.

There are many parties to most deals. Even the simplest, involving a programme 100% financed by a broadcaster envisaged for a single broadcast, will require a number of parties on the broadcaster's side, from scheduling to consulting the interests of other departments.

STAGES OF PRODUCTION

Pre-production

Pre-production is all about planning and anticipation. This is expressed in two formats: the budget, which anticipates spending, and the schedule, which anticipates the use of resources over time. These both rely on further plans. For documentary planning will typically be the result of a period of research; for drama, a period of scriptwriting. Even as writing and research are taking place, the idea that was agreed with the broadcaster is analysed into its component parts, and each is brought into line in an overall plan and budget. An outline budget will have been drawn up during the negotiation of the finance.

In pre-production a detailed budget will be developed which allocates resources to the various departments and phases of production: so much for the cast, so much for archive footage, so much for music, so much for editing, so much for digital effects etc. A budget will be set initially within broad outlines of the prices that can reasonably be expected from an up-to-date knowledge of the market. As pre-production continues, as scripts and/or research become more definite, these prices are progressively negotiated according to the detailed expectations of the production planning and the actual circumstances and quoted prices of suppliers. The precise outcome is only known, however, once the production is concluded. All the time that the production is active, the budget will be revised according to the actual, rather than predicted, expenditure.

The budget is refined as the detailed planning of the production continues. A documentary idea is researched to refine its details, find the most appropriate interviewees or (increasingly in contemporary documentaries) real people who can perform being themselves. Drama scripts are broken down into lists of locations and which scenes are to be shot in them. Locations are found and booked for the estimated number of days that it will take to shoot those scenes. Cast are found for the roles that were not specified for named stars in the contract.

A schedule emerges from the process, whose precise form is dictated by a network of other commitments (dates of star availability, limitations of locations etc), all geared to the contracted delivery date agreed with the broadcaster and other financiers. The larger the production (a 26 part drama series for example), the more complex the process. As part of the planning of the detailed schedule, prices for services are negotiated within overall guidelines of the budget. So there is a constant interaction between the two planning mechanisms: the schedule and the budget.

Pre-production is also a period of contract negotiation and writing. The detailed terms of the contract with financiers may still be in negotiation; certainly it is the period for negotiation of the terms under which the production company engages artists and technicians. It is also the period in which the underlying intellectual property rights ("copyright") in the production are contracted. Components such as the scripts, the original story, the music, the archive footage and so on all exist as separate artistic works and are protected by copyright. Copyright law differs in different countries. Production companies have to buy all or part of those rights.

The different rights in a piece of music, for instance, involve the simple right to incorporate the music in the programme (and that can be further limited to specific territories only) to a major or even exclusive interest in selling the music through other media such as CDs, radio sales etc. Music copyright is also divided into the copyright for the composition and the copyright for the performance. So the situation can quickly become complex because many copyrights ('underlying rights') are incorporated into even relatively simple programmes. This web of copyright is essentially another aspect of planning and anticipation. Copyright negotiations are usually governed by "what if" (as in "what if the programme is really successful"; "what if it is sold to a US network" etc).

Where extensive use of computer generated imagery is to be used, pre-production is also a period of image design and planning, particularly where images are to be shot to be combined into composites in post-production. Particularly important in television is the planning of the graphic style of a programme format, integrating everything from sets to title sequences, name captions and spin-off merchandise. Even modest productions now design a logo and an overall style as perhaps their prime means of differentiation in a crowded market. So it is important to see pre-production also as a phase of publicity. In a fast-moving market, the more publicity a programme can receive, the better.

The Shoot

Shooting, the creation and collection of images and sounds, takes many forms in television. Both film and video are used for different purposes. Considerable changes have taken place in the technologies available, and this has had visible effects on the kinds of programmes produced. The industry is also making increasing use of various forms of computer generated imagery and digital effects. Some genres use the specifically televisual form of the multiple-camera studio shoot, particularly game shows, daytime talk shows and sitcoms.

Studio production is intensive and involves a major division of labour. The director is physically separate from the action on the studio floor, and is mainly concerned with the 'instant editing' of the material from a number of cameras. Except in the case of live transmissions, this creates a master tape that can be trimmed and tightened later, especially when a subsidiary tape is made of a camera feed that is not being used in the live assembly. This is now a standard practice. The director is usually supported by a team who communicate through talkback with presenters and with the floor manager who is responsible for managing the activity in the studio space during recording.

The television studio is now a familiar feature of the self-reflexivity of the medium, and representations of it tend to emphasise its artificiality and the degree to which physical separation determines control over the finished product. For those who work within it however, studio practice is highly routinised and relatively inflexible, though not without its moments of tension. The multi-camera studio format is also adapted to location work, particularly in sport broadcasting where the liveness of the sport event is crucial. In these circumstances, the apparatus of the studio gallery or control room, as it is called in the US, is contained within an outside broadcast ("OB") truck, receiving the input from cameras stationed for the best possible points of view compatible with basic editing conventions (football and baseball are shot from one side of the field, for example).

Studio production brings all the elements of the programme together into one space and time, ensuring a contemporaneous and spontaneous atmosphere to the programme, which is a significant aspect of the genres that most use it. Other forms of shooting are more spread out over time, and imply a greater role for post-production. The technologies for both non-studio shooting and post-production have transformed fundamentally since broadcast television became a cultural commonplace. Television relied on celluloid film for the production of any recorded and post-edited material until the introduction of video tape around the beginning of the 1960s - almost a quarter of a century after the first regular broadcast TV services.

Film remained the preferred medium for larger budget production and is still used for much higher budget drama today. There are many reasons for this. Until digital technologies evened out the differences, film provided a superior image quality. In shooting conditions it can deal with a wider range of light levels within one shot, and is regarded as having a less 'flat' and 'bright' image quality. Film has historically seemed a more stable storage medium than tape, and its superior image quality meant that production on film could be more easily transferred to new formats like widescreen or digital storage. So the decision about shooting on film or tape has usually been linked to the planned quality and durability of the programmes.

Videotape, however, has advantages, especially in documentary situations where tape reels and cassettes have a larger capacity than film magazines. A filmed interview on 16mm had to break off every ten to twelve minutes to change the roll of film in the camera magazine; tape formats with a running time of an hour are commonplace. Until the introduction of cassette-based formats like Betacam in the mid-1980s, however, film technology was significantly lighter, more compact and more trustworthy in difficult conditions than tape. Betacam enabled high-quality location video shooting using two person crews similar to 16mm film crews. With the development of the camcorder in the mid 1990s, tape became significantly lighter and usable by single operators, often directors. This led to a revolution in documentary film-making, allowing longer and more reactive shooting periods and a more conversational and confessional style of interviewing.

In general, the technological aspects of shooting have simplified during the history of television. Fewer individuals are required and processes have become more automated. However, the human or 'front of camera' aspects of shooting have become more complicated. In fiction, there is a marked tendency to higher production values, with broadcast series like ER or NYPD Blue having distinctly filmic characteristics: complex sets, showy camera work, deep shots with plenty of background action. In documentary, it has become more difficult to negotiate the permissions to shoot as companies and ordinary individuals alike have become more used to the processes of television. Media awareness training is now a commonplace for large corporations and political campaigns; suspicion of television is also a deep rooted public attitude.

Post-production

In post-production, images and sounds are processed into the final programme form. In addition to the assemblage of images and sound from the shoot phase, post-production involves the creation and addition of material, such as commentaries, music and animation effects. This is an area of increasing sophistication, especially when a major amount of graphic design is involved. The demands of scheduling slots mean that programme durations have to be finely calculated, and the differing demands of broadcasters described above lead to a considerable activity of revision (or 'reversioning') of material to fit their formats and cultural demands.

All of this has been considerably helped by a technological revolution: the move from linear to non-linear editing. This has revolutionised the look and the internal pacing of programmes. Studio shot material is now routinely edited to sharpen its timing and to insert more complex visual information. Studio shows from the late 1980s, before the general use of non-linear editing, now seem archaic in their pacing and construction. Documentary and factual programming have seen an increasingly fast pace of shot transition and a sophistication in sound construction.

Non-linear is the term normally used for digital editing, using computer-based programmes such as AVID. Versions of these are now available for home PC use, and the key to the revolution is the increase in affordable computer memory. Video images are complex and require gigabytes of storage space, especially if all the information recorded on the tape is to be retained ('compression' refers to a process of averaging out image information to reduce the space it takes up). Tape is still used as a shooting medium in most circumstances because it is able to store large amounts of electronic data and is relatively robust. Some cameras have been developed using hard drives, but it is expected that tape will be eliminated as a shooting medium only when solid state cameras are developed which store information on chips. Before the introduction of multi-gigabyte hard drives, tape was used for editing as well. Computerised editing consisted of the precise copying of video material from one source tape to a master tape, one shot at a time in linear order, inserting effects by using an inline mixer.

Linear editing has a number of weaknesses. Any mistake involving a wrong shot length (omitted or unwanted material for example) means that the whole programme has to be reassembled on the master tape from that point on. Tape to tape editing involves a loss of image quality each time a transfer takes place ('generation loss'). So there are problems with the simple expedient of copying the rest of the programme to another tape, inserting the missing material, and then recopying the wanted rest of the programme to the master. The image quality of the end section has gone through two generations, so will be of inferior image quality. Three generations were deemed acceptable in most industrial contexts, so if the initial edit had involved a couple of generations, then the repair job would become more complex. The problem of generation loss was also a major constraint on the generation of special effects such as layered images and complex graphics. Non-linear editing has made these constraints a thing of the past.

Non-linear editing can take place at different levels of image quality depending on how much the image and sound information is compressed. So normal video editing practice is still to use higher rate of compression to make the initial edit choices and to assemble the programme in what is called an offline edit. The online edit then follows this template at the minimum level of compression. This process is dictated simply by expense: there is no point in tying up expensive equipment for long periods on one project during the period of assembly. An offline edit is usually calculated in days or weeks; an online in hours. This reflects a division of resources and labour from the practices of tape editing, where low-quality assemblies on VHS or similar formats enabled editors and directors to make creative decisions ("try things out") without the pressure of costs implied by an online edit.

Film editing follows a similar pattern, in which positive print material of rushes is assembled by the editor, and the finished cut is then sent to specialists who cut the negative material according to its guidance. Film editing was often preferred to linear tape because of its greater flexibility, but the non-linear revolution has largely replaced traditional editing on celluloid even within the feature film business. Non-linear editing also enables a level of sound post-production that was difficult to attain with linear tape, so that many productions now involve considerable work of sound design and a separate stage of sound mixing after the images have been assembled in their final form. Non-linear editing has also enabled the creation of new kinds of job in the industry, such as the sound designer or the colourist, as well as eliminating others such as the assistant editor.

Post-production also involves considerable administrative work. The final accounts are produced, and in productions that are going badly, this often involves an attempt to reduce the costs of post-production. Paperwork is produced in large quantities, describing the programmes as commodities in various ways: everything from the transcripts of dialogue needed for subtitling and translation, to detailed descriptions of the origins of all the footage used and any copyright restrictions, to contracts with key personnel and even publicity material.

Marketing

Marketing is an integral part of broadcast production. During pre-production and shooting various kinds of publicity material are prepared: plot synopses, stills, interviews with leading actors, etc. Post-production involves the production of trailers and other promotional material from the shot footage. Sometimes this is the responsibility of the production company; sometimes it is undertaken by broadcasters themselves. There is a particular challenge to marketing a show on broadcast television: there is a very short span of time (usually only a week or two) to publicize the one-time transmission of a show or an episode of a show at a given time on a given date. In the competitive television market, channels are spending increasing amounts of on-screen time and production resources in producing on-screen promotional material for forthcoming shows, to the extent of shooting expensive material simply for this purpose.

Marketing also involves attempting to control media coverage. The horizontal integration of media companies has greatly affected the marketing of TV programs. The Disney Channel, for example, routinely cross-promotes TV programs, Disney-produced or distributed movies, computer games, theme parks, toys, etc. The News Corporation in Britain uses its newspapers (the Sun and The Times) to promote programmes and events on its BskyB satellite channels. Marketing has also been extended to the web, with banners at the bottom of the screen driving viewers to program-linked websites and, conversely, websites set up especially to promote a new show or serve as a vehicle for fan interaction.

Marketing is not always a matter of spreading information, however. In 1985, the BBC decided that the launch of its new primetime soap Eastenders was to be shrouded in secrecy, with no press access to contracted actors and no advance leaking of the plot. Even now that the series is well established, the feeding of information to the press is still tightly controlled, so that the information that an actor is leaving the cast will trigger speculation about the way in which this will be managed in script terms, enhancing public interest in the series. Elsewhere, gossip about the cast and their views on the characters they play are provided to magazines aimed at the core sections of the audience, primarily pre-teen and early teens.

Such are the marketing activities involved in production for broadcast television. There is, however, a lot more to television than broadcasting, and a lot more to production for broadcasting than just programmes and ancillary material. The rest of this chapter outlines some of the major aspects of television beyond the cultural dominance of broadcast programming.

BEYOND BROADCAST

Broadcasting as defined here is a culturally dominant way of using television technology. Other forms exist which dwarf broadcasting in terms of their economic importance in production terms. Some, like the use of surveillance cameras in public spaces, involve little in terms of production activity until crucial images require enhancement for detection purposes. Others, currently represent relatively small areas of activity, but are the harbingers of a trend: the application of television technologies to everyday commercial activities.

This section will deal briefly with four areas: the production of commercials for television; the use of television in corporate communication; video art and the relatively new domain of personal production.

Commercials

Much television production outside of the broadcasting industry itself is concentrated in advertising. Indeed, advertising pays for much of the production activity of the broadcasting industry as well. Spot advertising underpins much of the current broadcasting industry, yet broadcasting is simply one way of reaching consumers, and, increasingly, it is seen as a relatively ineffective means of doing so. When a major brand like Heinz decides to shift resources out of television advertising into other forms of influencing consumers, the whole broadcast industry indulges in a bout of severe anxiety. Broadcasting, though culturally important, is a relatively minor industrial activity. As a British investment advisor once put it: "they're terribly nice people to deal with, but you could buy the whole industry out of the retained profits of ICI".

Many of the concerns of commissioning practice and editorial niceties that I have described simply do not exist for advertising, so advertising production should be considered as a separate form of production, as it is in the audio-visual industries. The genesis of a commercial is very different from that of a television programme: the concern is the message and the brand, the impact in the market. Creative decisions are guided by these concerns and are made by more disparate groups of people than is commonly the case in broadcasting. They will involve campaign directors as well as film directors;

market researchers as well as programme researchers; clients as well as creatives. Advertising involves a number of different functions, any or all of which can be found within one corporation.

The first is the client, the company with a product to sell that has already been the subject of considerable consumer research and consequent conceptualisation. Second is the agency that conceptualises the advertisement (the concept). Third is the production house that produces the physical objects, advertisements, which often exist in related forms across a number of different media. Fourth is the organisation that buys the display space for the advertisement, the spot in the advertising break or the programme sponsorship deal. The common conception of the 'advertising agency' is of a company that undertakes all these functions for a client, including a considerable amount of the initial market research and product conceptualisation. The situation is often far more fluid, with large corporations taking various functions 'in house' or seeking to control the budgetary aspects of the physical production of commercials by imposing their own production controllers on separate production companies.

Production is dominated by a perfectionism that is rare in broadcasting. Since every frame of an advertisement will be seen dozens of times by an average viewer, every implication of every image has to be teased out. Every frame of a high-budget advertisement is meticulously pre-planned, and will freely borrow techniques from the film industry including the lavish use of computer generated imagery. The production cost per minute of the average advertising break will be many times that of the broadcast programming that separates one break from another. Yet there also exist commercials whose skimpy production costs seem entirely out of keeping with the cost of putting them on air.

One of the great points of tension within television production is the relationship with the advertising industry. Spot advertising was devised as the simplest means of separating the interests of the corporations using broadcast TV for advertising from those of the corporations making money from broadcasting itself. So two separate but inseparable interests have been at work within broadcasting since the start. Even the early BBC, whose public service remit rigorously excluded advertising, had agonised discussion about how to deal with brand names that were in common circulation ('Hoover' for vacuum cleaner etc.), and what details they should give of commercial gramophone records that they played.

In the movement away from the dominance of network television and the fragmentation of the television audience across dozens if not hundreds of channels, spot advertising has become less effective. Instead, sponsorship has increased, seeking to associate commodities with programmes on the basis of a connection between their brands. The confectionery company Cadbury's sponsors ITV's long-running soap Coronation Street with the legend 'brought to you in association with Cadbury's Drinking Chocolate', linking a familiar and comfortable programme with a product that hopes to produce the same feelings in its consumers. The spot commercial, however, has found a multitude of other outlets. Public display television monitors can be found at railway stations, gas/petrol stations, supermarkets, planes, in short they are in all the waiting areas that are part of modern existence. Video images in motion passes across them, and given the nature of the spaces involved, sound can often not accompany them. Many kinds of spot commercial are ideal for this visual form of display.

Corporate Communication

Global corporations rely on television to create a corporate culture with a sense of common purpose, using it in several distinct ways. Management uses television as a means of direct and charismatic communication, and television training is a key part of the skills of a senior manager. Live feeds of important addresses from senior managers are the most effective means of communication at times of crisis or celebration. The preparation for the launch of a new product involves high budget productions communicating the virtues of new products in ways that are distinct from the presentation of that product to customers.

With the increasing complexity of the modern workplace, video is used routinely to provide instructional material detailing workaday processes. As workplaces become screen-based rather than paper-based, this kind of visual instructional material will increase in importance and complexity. Some of these forms draw upon broadcasting, if only to enhance their cultural status. Corporate communications will often use television faces or voices as presenters, and sometimes use real or imitated TV formats to get their message across.

Corporate communication will grow as television becomes a more usual means of spreading information. Already corporate websites use video quite routinely, and the building industry increasingly uses video 'walk-throughs' of planned buildings to secure permissions and to make advance sales. Estate agency/realtor businesses use video material of homes for sale as a means of aiding customer selection.

Video Art

Artists have used video since the late 1960s, when Sony introduced the Portapak, using an open reel to reel tape rather than a cassette (cassettes were introduced in 1973). Video provided artists with the ability to capture time and process much more easily than did film, enabling collaborative work with performers. Artists like Bill Viola also worked with the physical nature of the image display itself, placing monitors under water (a recurrent theme in his work) or on their sides which suddenly made them look surprisingly tall and narrow. Video art often plays with the incongruity of this domestic form within the gallery setting, with artists like Gillian Wearing taking the television interview relationship into areas unexplored by broadcasting.

Her first major work was Signs that say what you want them to say and not Signs that say what someone else wants you to say (1992-93) in which she photographed random passers-by holding statements that they had written down spontaneously. In one image a smartly-dressed young businessman with a mild expression holds up a sign saying simply, 'I'm desperate'. Television technology presents interesting challenges for the workings of the art market, with digital technology eroding the uniqueness of the artistic work even further than still photography. Wearing's solution is ingenious: she sells a limited number of tapes, each with a playback machine and monitor. These together constitute the artwork. Others have used television to create webcam events, merging broadcasting with artistic concerns with time, process, biography and exhibitionism to create a new category of artistic experience.

Personal Production

The most fundamental change in television within the recent past has been in the availability of digital production tools for the ordinary domestic consumer. Digital camcorders can be combined with editing software on domestic PCs to provide an effective production route. Dissemination can take place through websites as well as video cassettes. The real significance lies in the increase in audio-visual communication that it enables. Video communication can replace written communication where appropriate. Personal production also provides a growth in the audio-visual literacy to the general population, which feeds back very quickly into production for broadcasting and for advertising. The result is not a democratisation of broadcasting so much as the introduction of the audio-visual into person-to-person communication. Broadcast television may well be the cultural dominant at the moment, but in production terms it is already a specialised area, and it will become an exceptional model of production as new forms proliferate.

THE MAKING OF TELEVISION ENTERTAINMENT

The dialectic of cultural assertion and hegemonic incorporation is well illustrated by the production of American NETWORK television entertainment, which shall be the principal subject of the rest of this essay. In one way, TV entertainment imprints its content with hegemonic ideology, aiming to organize the largest possible consuming audience and to work a hegemonic effect on it. The result is generally what the industry calls LOP: Least Objectionable Programming. Yet in another way, TV entertainment enfolds and amplifies centrifugal tendencies in the culture. Television critics have been heard to complain that television shows are insipid and slovenly.

True enough; assembly-line production schedules are not kind to quality. But much of TV's insipidity may be understood as a patterned ambiguity. If, from an aesthetic point of view, simplistic ambiguities amount to equivocations, from a sociological point of view they are perfectly tailored to appeal to the components of a complex society. The aesthetic and political blandness of most TV entertainment results not from the technology of television as such - not from the size of the screen, not from the texture of the image - nor from a shortage of money for production, nor from the audience's presumed lack of intelligence, but from television's dominating social purpose.

NETWORKs in the United States, and state-controlled TV organizations in Europe, set out to communicate to mass audiences. But mass audiences contain a myriad of subcultures, each inclined (though not unambiguously) toward a distinct (though not autonomous) system for decoding, inflecting, stretching the culture's central symbols. Contrary to any vulgar Marxism or ethnographic relativism that might read the culture as nothing but a field of essential conflict over premises, feelings, and values, corporate capitalist societies have certain collective representations: this is the truth in Durkheimian and Tocquevillian assessments of modern culture. The function of mythic images is to deny social cleavage and to formulate a collective identity for a fragmented society. (What conservative tradition cannot explain, though, is the nature and source of the unifying myth.

Why capitalist hegemony, and who gets to articulate its symbolic forms?) Not least amidst the universal grounding of contemporary culture is the high value placed on images themselves; the realm of spectacle emerges, conditioning experience and penetrating it. Still, the society as a whole is no homogeneous whole: despite tendencies in that direction, it is also culturally polyglot. Fragments of folk culture remain, regionally and ethnically distinct. And distinct groups selectively appropriate and construe what they find in the larger cultural domain; new formations arise and position themselves vis-à-vis others; subcultures conflict, transform themselves, and sometimes decay.

Amidst this flux, the culture industry cannot take it for granted that its audiences will read its products unambiguously. One consequence is that production is dominated by the LOP formula. And the shows' susceptibility to multiple interpretation becomes the fulcrum of production strategy in a second way as well: television, far from fleeing ambiguity, makes possible a certain range of counterposed, discrepant, and even oppositional interpretations.

How people will read a program is not fully predictable. Entertainment taken as a whole, and even, at its most sophisticated, show by show, can thus please several audiences at once: its messages may run with more than one grain. How can this happen? The denotations of visual signs are relatively unambiguous: the image of a man wearing a white hat and riding a horse stands for a real man riding a real horse.

But connotations vary: the image may stand for a heroic man-to-the-rescue, a countercultural rebel, or a reprehensible imperialist, depending on who is watching. The first reading is dominant and conventional; the second might be alternative; the third, oppositional. Connotations will depend on group identity and ideology, as well as individual cognition and circumstance. The point is that multiple connotations coexist in the artifact; they are layered into it, although rarely in a premeditated way, from the start. Bundled together, they permit different segments of the audience to partake of the same artifact yet not quite the same experience; they permit the audience to decompose the show differently, to take more or less account of different parts and characters, and to draw different conclusions - while tending to take the show's hegemonic definition of the situation as a fixed point of departure, an imposing symbolic universe.

Varying 'effects' amidst hegemony result from the fact that the people who watch television are simultaneously:

- Members of a mass audience, rendering the show popular, binding themselves in webs of contemporaneous though invisible relations to others as serialized consumers;
- Members of families and subcultural groups who take the show in and digest it through their respective interpretive filters; and
- Individuals imposing their personal interpretations on the shows, and perhaps insisting on their individuality by thinking or saying, 'I don't know what anybody else thinks of it, but I think...' The mass audience is not only the arithmetic sum of disconnected individuals, but a skein of social clumps.

Against this background, the packaging process is no simple matter of élites fastening on their purposes and generating material to suit, as if the media of communication were simply spigots. Amidst a welter of social forces, many choices are overdetermined. Others emerge from a conflict among forces, and the outcomes are contingent: they might have turned out differently, even given the organizational structure of the culture industry.

To account for the emergence and shape of any given program, a comprehensive analysis needs to take account of a range of factors:

- *The organizational structures of the production process*: Specifically, among NETWORKs, advertisers and advertising agencies, and producers. NETWORKs retain determining power because they control access to the major markets; beneath them, the major producers have oligopoloistic power. One filter controls access to the one beneath it. The NETWORKs commission pilots, and then choose from among them. Given the dominion of professional standards, the costs of production are so great as to give the NETWORKs and major production studios (which occasionally produce 'independently' for syndication as well) enormous power over what is finally broadcast. In the words of Bob Daly, President of CBS's entertainment division, 'We get 2000 ideas for series every year. Out of that 2000, we commission 200 scripts, we make 40 "pilots" so that we can look at them to decide if they should go out on the air. Of those 40 pilots, only eight actually become series and get shown. And of those eight - only three actually survive one season'. To explain the three survivors, of course, we must have recourse to audience mentalities.

- *The NETWORKs' market strategies*: The NETWORKs aim to garner the maximum attention from that part of the total possible audience which spends the most money. Thus, other things being equal, a show attracting a given share of the 18-to-49-year-old audience, as measured by the A. C. Nielsen Company, will succeed more than a show attracting the identical share of an older audience. Increasingly sophisticated 'demographic' measurements enable the NETWORKs to make increasingly specific market judgments. The demise of Westerns may be traced partly to the fact that their major audience is disproportionately older and disproportionately rural: not the audience most capable of high-volume consuming.
- *The values, beliefs, and strategies of popular-cultural elites*: NETWORK executives and producers, and then, secondarily, writers and even actors. Class identifications, shifting aesthetic and political ideals and ideological tolerances, the producers' and packagers' and writers' sense of 'what's in the air,' 'what might fly,' all play a part in generating ideas for pilots. But story ideas are neither born free nor promoted equally; they must appeal to the elite of the production process or they cannot proceed to the stage of production. They do not get a chance to find their audience unless they filter through all the layers of the corporate choice system. Moreover, despite the obligatory references to individual talent in the industry, individuals succeed only in so far as they match the corporation's expectations.
- *Specific corporate habits and practices*: I refer here to contingencies: factors in corporate choice which stand relatively free of the deeper structural constraints and strategies. The positioning of a show has a great deal to do with its ratings, since people tend to watch television continuously and would rather not change channels unless impelled to do so. Thus, other things being equal, a show that appears just after a high-rated show will do better than the same show appearing just after a lower-rated one. A show that appears opposite a high-rated show on another NETWORK will do less well than if it appeared opposite lower-rated shows. The NETWORKs' maneuvering their shows in relation to others can get ferocious. Another corporate habit is a particular adaptation to the

NETWORKs' competitiveness: when a particular show succeeds, the NETWORKs tend to reproduce what they deem to be its salient features. The success of a cop show generates other cop shows: Charlie's Angels generates 'T & A' clones (to use one industry term: 'jiggle shows' is another). The pendulum generally swings too far: audience interest in a given formula does not extend to cheap imitations, and the market for the formula is oversaturated within a season or two.

- *Audience mentalities*: Audience do not express their preferences directly; what audience ratings measure is not what people might abstractly prefer in a world they might imagine, but what they will at least tolerate, at most like, given what television offers and what else they might do with their time. Producers have only shadowy images of what audiences want. They reconnoiter. They rely on their immediate social circles, and the ideas that gravitate their way (partly affected, of course, by new entries' ideas of what is commercial), and on the versions of popular belief that filter into their life-worlds - through mass media, ironically enough. Audience mentalities do affect programming, but indirectly; the culture industry does not passively reflect audience desires; but refracts them, actively and darkly, through its organizational and ideological glasses.

So much for the abstract, a priori categories of explanation. How particular shows emerge, having percolated through this set of filters, must be studied empirically. Mindful of the appalling fact that there is not a single book-length sociologically-informed study of American television's ideological content, I am about to launch into such an investigation. I will be looking to analyze an entire season's array of entertainment series, interpreting their contents, looking to explain which pilots and which scripts do and don't get produced, and which series do and don't last. By interviewing producers and studying television archives, I hope to understand the relative weights of the above-mentioned shaping forces.

Before I go at interpreting the material with the care it deserves, and plumbing specific audience responses to the shows, it makes no sense to claim too much for an assessment of the meanings that TV entertainment carries. But at this early stage I do want to set out a sketch, a rough prospectus, enumerating a number of elements of the television discourse, conventions in which ideological hegemony is embedded: format; plot formula; genre; setting; character type; images of social and psychological conflict and its solution; images of authority, the State, family, work, and social movements; images of emotion, its texture and legitimacy. Consider the following discussion a sort of prospectus for the more detailed, more systematic, more historically grounded and socially situated study to come.

Format

Until recently at least, the TV schedule has been dominated by standard lengths and cadences, standardized packages of TV entertainment appearing, as the announcers used to say, "same time, Same station.,, This week-to-weekness-or, in the case of soap operas, day-to-dayness-obstructed the development of characters; the primary characters had to be preserved intact for next week's show. Perry Mason was Perry Mason, once and for all; watching the reruns, only devotees could know from character or set whether they were watching the first or the last in the series. For commercial and production reasons, which are in practice inseparable-and this is why ideological hegemony is not directly reducible to the economic interests of elites-the regular schedule prefers the repeatable formula: it is far easier for production companies to hire writers to write for standardized, static characters than for characters who develop, Assembly-line production works through regularity of time slot, of duration, and of character to convey images of social steadiness: come what may, "Gunsmoke" or "Kojak" will occupy a certain time on a certain evening. Should they lose ratings (at least at the "upscale" reaches of the demographics, where ratings translate into disposable dollars), their replacements would be-for a time, at least!-equally reliable. Moreover, the standard curve of narrative action-stock characters show their standard stuff; the plot resolves-over twenty-two or fifty minutes is itself a source of rigidity and forced regularity.

In these ways, the usual programs are performances that rehearse social fixity: they express and cement the obduracy of a social world impervious to substantial change. Yet at the same time there are signs of routine obsolescence, as hunks of last year's regular schedule drop from sight only to be supplanted by this season's attractions. (The very concept of "season"-in TV as in fashion, and in the opera, ballet, and theater of high culture-claims the authority and normality of nature, s cycles for man-made products.) Standardization and the likelihood of evanescence are curiously linked: they match the intertwined processes of commodity production, predictability, and obsolescence in a high-consumption capitalist society.

I speculate that they help confirm audiences in their sense of the rightness and naturalness of a world that, in only apparent paradox, regularly requires an irregularity, an unreliability, which it calls progress. In this way, the regular model changes in TV programs, like the regular changes in auto design and the regular elections of public officials, seem to affirm the sovereignty of the audience while keeping deep alternatives off the agenda. Elite authority and the illusion of consumer choice are affirmed at once-this is one of the central operations of the hegemonic liberal capitalist ideology.

Then, too, by organizing the "free time" of persons into end-to-end interchangeable units, broadcasting extends, and harmonizes with, the industrialization of time. Media time and school time, with their equivalent units and curves of action, mirror the time of clocked labor and reinforce the seeming naturalness of clock time. Anyone who reads Harry Braverman's Labor and Monopoly Capital can trace the steady degradation of the work process, both white and blue collar, through the twentieth century, even if Braverman has exaggerated the extent of the process by focusing on managerial strategies more than on actual work processes. Something similar has happened in other life-sectors.

Leisure is industrialized, duration is homogenized, even excitement is routinized, and the standard repeated TV format is an important component of the process. And typically, too, capitalism provides relief from these confines for its more favored citizens, those who can afford to buy their way out of the standardized social reality that capitalism produces. Beginning in the late 1970s, the home videocassette recorder enabled upscale consumers to tape programs they would otherwise have missed-I by 1984 there were ten million in American homes. The widely felt need to overcome assembly-line "leisure" time thus became the source of a new market-to sell the means for private, commoditized solutions to the time jam.

Commercials, of course, are also major features of the regular TV format. There can be no question but that commercials have a good deal to do with shaping and maintaining markets-no advertiser dreams of cutting advertising costs as long as the competition is still on the air. But commercials also have important indirect consequences on the contours of consciousness overall: they get us accustomed to thinking of ourselves and behaving as a market rather than a public, as consumers rather than producers or citizens. Public problems (like air pollution) are to be understood as susceptible to private commodity solutions (like eyedrops). In the process, whether or not we are offended or annoyed by commercials, they acculturate us to interruption through the rest of our lives.

Time and attention are not one's own; the corporations in effect advertise their own dominion along with their products. Regardless of the commercial's effect on our behavior, we consent to its domination of the public space. Yet we should note that this colonizing process does not actually require commercials, as long as it can form discrete packages of ideological content that call forth discontinuous responses in the audience. Even public broadcasting's children's shows take over the commercial forms by herky-jerky bustle. The producers of "Sesame Street," in likening knowledge to commercial products ("and now a message from the letter B"), may well be legitimizing the commercial form in its discontinuity and its invasiveness. Again, regularity and discontinuity, superficially discrepant, may be linked at a deep level of meaning. And perhaps the deepest privatizing function of television, its most powerful impact on public life, may lie in the most obvious thing about it: we receive the images in the privacy of our living rooms, making public discourse and response difficult. At the same time, the paradox is that at any given time many viewers are receiving images that do not accord with many of their beliefs, thus challenging their received opinions.

Television routines have been built into the broadcast schedule since its inception. But arguably their regularity has been waning since Norman Lear's first comedy, "All in the Family," made its NETWORK debut in 1971. Lear's contribution to TV content was obvious: where previous shows might have made passing reference to social conflicts, Lear brought wrenching social issues into the center of his plots. Lear also let his characters develop. (Previously, only the children in family series had been permitted to change-a forced maturation.) Edith Bunker grew less sappy and more feminist and commonsensical; Gloria and Mike moved next door, and finally to California. On the threshold of this generational rupture, Mike broke through his stereotype by expressing affection for Archie, and Archie, oh-so-reluctantly, but definitely for all that, hugged back and broke through his own.

Other Lear characters, the Jeffersons and Maude, had earlier been spun off into their own shows. (Since Lear's success with spin-offs, popular actors on other shows have been able to bargain themselves into their own series "Rhoda" from "The Mary Tyler Moore Show," "Flo" from "Alice," ad infintum.) Lear's precedents have flourished; they were built on intelligent business perceptions that an audience did exist for situation comedies that directly addressed racism, sexism, and the instability of conventional families. But there is no such thing as a strictly economic explanation for production choice, since the success of a show-despite market research-is not foreordained. In the context of my argument, the importance of such developments ties in their partial break with the established, static formulae of prime-time television.

Daytime soap operas and their prime-time variants have also been sliding into character development and a direct exploitation of divisive social issues, rather than going on constructing a race-free, class-free, feminism-free world. And more conspicuously, the "miniseries" has now occasionally disrupted the takenfor-granted repetitiveness of the prime-time format. Both content and form mattered to the commercial success of "Roots"; certainly the industry, speaking through trade journals, was convinced that the phenomenon was rooted in the break the series made with the week-to-week format. When the programming wizards at ABC decided to put the show on for eight straight nights, they were also, inadvertently, making it possible for characters to develop within the bounds of a single show. And of course they were rendering the whole sequence immensely more powerful than if it had been diffused over eight weeks.

The very format was testimony to the fact that history takes place as a continuing process in which people grow up, have children, die; that people experience their lives within the domain of social institutions. This is no small achievement in a country that routinely denies the continuity and directionality of history.

Plot Formula and Genre

The conventions of television entertainment are flexible precisely because they operate under limited but real market constraints: they enable popular forms to express and manage shifts in the available stock of ideology. In the 1950s, the NETWORKs tended to reproduce the image of a society at one with itself, without significant social tensions-though even then, the shows sometimes did register a muted sense of the routine frustrations caused women by male supremacy ("I Love Lucy") and the routine psychic injuries done to workers ("The Honeymooners"). For the most part, the world of television was what Herbert Gold called a world of "happy people with happy problems."

Then, after social conflict grew explosive in the 1960s, television began, gingerly and selectively, to incorporate certain symbols of dissonance and changing "life-styles "-racial and ethnic consciousness, hip professionalism, new living arrangements, ecological awareness, and sanitized middle-class versions of feminism. Where the family dramas and sitcoms of the 1950s usually denied the existence of deep social problems in the world outside the set, or sublimated them into obscurity, programs of the 1970s much more often acknowledged that the world is troubled and problematic, and then proceeded to show how the troubles could be domesticated. From "Ozzie and Harriet" and "Father Knows Best" to "All in the Family" and "The Jeffersons" marks a distinct shift in formula, character, and slant: a shift, among other things, in the image of what the larger social world amounts to, how stable it is, and how a family copes with it.

Some shows become popular by speaking directly to a compact, socially homogeneous public. But more likely-or so it seems on the near side of systematic research-the most popular shows are those that succeed in speaking simultaneously to audiences that diverge in social class, race, gender, region, and ideology: and this because of the mass market imperative of NETWORK television. To package the largest possible audience, the NETWORKs must offer entertainment that is literally broadcast: appealing to a multiplicity of social types at once.

It way embody its values directly or indirectly, overtly or covertly, but it will do best if it embodies them ambiguously enough to attract a variety of audiences at once. The "socially relevant" situation comedies produced by Norman Lear, beginning with "All in the Family in 1971, may well have broken through to ratings success precisely because they directly and ingeniously broached the divisive social issues of race and political culture. Some studies of audience response have suggested that "All in the Family" permitted audiences on both sides of the generational and political chasms to feel confirmed in their attitudes: older conservatives rooted for Archie, younger liberals for Mike. (One aspect of the show's hegemonic framing was that most of the divisive issues were represented as matters of generation, not class or power.) Why the great success of Lear's new genre? A brief excursus may suggest more general explanations for shifts in program formula. For one thing, the historical timing of a show bears heavily on its commercial prospects. ABC rejected "All in the Family" before CBS bought it. In contrast, consider an earlier attempt to bring problems of class, race, and poverty into the heart of television: CBS's 1963-1964 "East Side, West Side," in which George C. Scott played a caring social worker who was consistently unable to accomplish much for his clients however hard he tried. As time went on, the Scott character came to the conclusion that politics might accomplish what social work could not, and went to work as the assistant to a liberal Congressman. It was rumored that the hero was going to discover there, too, the limits of reformism-but the show was then canceled, presumably because of low ratings.

In the middle and late 1960s social conflict had been too inflammatory, too divisive, to permit NETWORK television to indulge in, Lear's formula for accommodation through ambiguity. And Lear's shows, by contrast to "East Side, West Side," have lasted partly because they are comedies. Audiences will partake of comedy's ready-made defenses to cope with threatening impulses, especially when the characters are, like Archie Bunker, ambiguous normative symbols. The comedy form allowed white racists to indulge themselves in Archie's rationalizations without seeing that the joke was on them. And finally, as Michael J. Arlen once pointed out, Lear was further inspired to unite his characters in a harshly funny ressentiment that was peculiarly appealing to audiences of the Nixon era and its cynical, disabused sequel. One alluring subtext of the show was that a family could hold together despite everything pressing on it from outside; the family could encompass the social conflicts that had seemed to be tearing the country apart.

So here we see the range of textual features that an interpretation can take into account. But how was the show possible in the first place? Structural and organizational explanations for the show's origins and success begin on the shoulders of the interpretive reading. Lear was in a position to shatter conventions, first of all, because producers had gained in the power to initiate. In the 1950s, sponsors directly developed their shows and were thus better able to control and sanitize content.

When the NETWORKs took much of that power away in 1960, in the wake of the quiz show scandal, they began to make decisions in the interest, as it were, 'Of advertisers in the aggregate. By the late 1960s, television had become so important an advertising medium that advertisers were standing in line to buy scarce commercial time. Thus NETWORKs were somewhat more willing to take chances with risky shows, knowing that if a few advertisers were offended there would likely be others eager to replace them, unless the content were in flagrant violation of hegemonic norms (the adventures of a union organizer, say, or a fundamentalist preacher). Changes in content also flow from changes in social values and sensibilities—changes among producers, writers, and other practitioners, but also changes they are aware of in the audiences that are most salient to them. Lear's own political position-he is a major contributor to liberal causes-was mostly beside the point: it mattered only insofar as it attuned him to a new marketing strategy. Lear was aware that there was, in the 1970s, a large audience nurtured ideologically in the opposition movements and counterculture of the 1960s and now preferring to acknowledge and domesticate social problems, hoping to reconcile contraries in imagination rather than to ignore or deflect them.

The whole texture of social life had changed: where the official mythology of the 1950s had stressed cultural consensus, the consensus had cracked in the 1960s and ideological divisions had surged into the open. And crucially, there were now writers available who had ideological roots in the opposition movements of the 1960s, though by themselves they could not account for content or success; there was also, after all, such a supply in the 1950s, but it was cut off by the blacklist, which exercised a chilling effect on subject matter and plot formula.

That chill had been produced by the NETWORKs, which accorded veto power to police agencies and professional associations (notably the American Medical Association), thus acting instrumentally in behalf of a hegemonic interest originating on the outside. ABC, for example, gave routine veto power to the FBI over its longrunning series of the same name. On one occasion, the TV writer David W. Rintels was asked to write an episode of "The FBI" on a subject of his choosing. Rintels proposed a fictionalized version of the 1963 bombing of a Birmingham church, in which four black girls were killed. Rintels wrote later:

"The producer checked with the sponsor, the Ford Motor Company, and with the FBI-every proposed show is cleared sequentially through the producing company, QM; the Federal Bureau of Investigation; the NETWORK, ABC; and the sponsor, Ford, and any of the four can veto any show for any reason, which it need not disclose-and reported back that they would be delighted to have me write about a church bombing subject only to these stipulations: the church must be in the North, there could be no Negroes involved, and the bombing could have nothing at all to do with civil rights.

After I said I wouldn't write that program, I asked if I could do a show on police brutality, also in the news at that time. Certainly, the answer came back, as long as the charge was trumped up, the policeman vindicated, and the man who brought the specious charge prosecuted." On another occasion, a NETWORK acted as direct guardian of the hegemonic limits: NBC refused to permit "Dr. Kildare" to run a show about venereal disease, this time despite clearance from the AMA, the National Education Association, and the surgeon general of the United States.

But times have changed. Medical shows on venereal disease have been aired. The FBI is no longer exalted. From 1977 through 1982 "Lou Grant" took account of contemporary issues from a muckraking point of view. The NETWORKs now arrogate to themselves the right and the power to legitimize the respectable framing of social problems, and they yield less authority than before to agencies of the State. Now, when a specific show slants to the right, it is most likely because the producers anticipate an audience in that direction; direct state intervention is not necessary.

By the 1970s, in short, the cocky, commercially booming NETWORKs had eclipsed specific advertisers and government agencies; they had made themselves the direct shapers of hegemonic content. And they had devised a new formula to coexist with some older ones and to displace others. Observing shifts in the tolerances and potential enthusiasms of the market-especially the younger, more liberal, more "Permissive" upscale market-they had downplayed crude censorship, and preferred to take account of shifts in social ideology by domesticating them, by offering hegemonic solutions to real problems.

This active processing of deviance and lifestyle seems automatic, the sum of innumerable production decisions played out as if by reflex and commercial instinct. Yet this hegemonic strategy at times surfaces into the thinking of culture industry elites: it may be quite sophisticated, quite precise in its implications for television content. In a March 1960 speech to the Southeast and Southwest councils of the American Association of Advertising Agencies, held at St. Thomas, Virgin Islands, Jane Fitzgibbon, senior vice-president of the research firm of Yankelovich, Skelly & White, told advertising executives that television's role has been "to legitimize new life styles after they have emerged," rather than to inaugurate them. "I think we can honestly say," she went on, "that the television medium has speeded up what you might call the filtration of new values and new life styles throughout the population.

Television does this simply by documenting new life styles, particularly through its news service." She took account of cultural conflict: "We now have at least two audiences to appeal to if we talk in grosso modo terms: traditional values-and they are still around-and new values. But within the new values segment there are two groups. There's the self-fulfillment, quality-of-life, self-improvement segment; and there's the experience and the escapist-oriented new values segment. What I see this leading to is more and more audience fragmentation, probably smaller market shares, smaller rating shares and probably few blockbuster shows." And finally, she nicely articulated the hegemonic managerial task for program developers:

"Television must be consistently attuned and alert to life-style changes (this goes for the advertisers as well as the writers), so that it can accurately and responsibly portray them at a point in time when the public will neither be bored because they are too outdated, nor outraged because they are too far out on the fringes. Instead, television's portrayal of societal change can insure that the public be stimulated, informed, sensitized, reassured about what is happening in their own personal lives and the lives of other people in the world at large." The "accurate and responsible portrayal" that forswears both the outdated and the far-out; the combination of stimulation, information, sensitization, and reassurance-these are the terms with which the strategy of domestication is accomplished.

Another example suggests both the subtlety and the possible intentionality of this process. The popularity of "Charlie's Angels," beginning in 1976, suggests that television producers have learned how to appeal to elements of the new feminism and to its opposition at the same time. The Angels are highly skilled, motivated, working women; they show a certain amount of initiative. How else would they appeal to a certain female audience, toward whom many of the show's ads are beamed? (They represent, paradoxically, the subordinated side of Mary Tyler Moore, the progenitor of the single-woman show, who signals that some sort of feminism is here to stay as a "new life style" for the single career woman.)

At the same time, plainly the Angels are sex objects for men, as the cults of Farrah Fawcett-Majors and Cheryl Ladd attest. And they are subordinated: they usually rely on Charlie's aid to bail them out of the dangers to which their spunk has exposed them, It is no small element of the show's appeal to men that Charlie, the detective boss, is never seen. Male authority is invisible, and the "girls" are kept free of romance. Thus, in the male viewer's fantasies, the Angels remain available-to him and him alone. He, in unconscious fantasy, is Charlie, ever supervising, ever needed, ever jovial, ever returned to. The show thus caters simultaneously to feminism and to backlash against it; it permits men to indulge prurience while psychologically admitting women's importance to the workforce.

(I must add that I had thought this for months, marveling at the impersonal ingenuity of television production process, when I came upon a quotation from an anonymous "top television executive of one of America's big three NETWORKs [who] said quite seriously, 'A series like Charlie's Angels performs a very important and valuable public service. Not only does it show women how to look beautiful and lead very exciting lives, but they still take their orders from a man'". This is not to say that such thinking preceded the show and its popularity. A post hoc theory of function is not a program for strategy. But it is interesting that savvy NETWORK executives may become conscious theorists of hegemony, may recognize it when they see it.

In general, then, genre is necessarily sensitive; in its rough outlines, if not in detail, it brews a blend of popular sentiments." Sometimes genre runs in advance of hegemonic ideology; more often it lags. A fine analysis of its themes will probably reveal elements of both in any given case. New genres coexist for a time with old ones, which may themselves be rooted in traditional forms: the western, the detective story, the variety show. New genres sometimes transpose old ones: the ever-popular "Star Trek," for example, was essentially a western whose team of professional lawkeepers operated in space. Without attempting here a thorough account of the metamorphoses of TV genre, I can suggest by way of hypothesis a few other signs of NETWORK sensitivity to actually and potentially shifting moods and group identities in the audience -One decisive clump of questions to be asked of television entertainment is: What is its attitude toward authority? Where does it locate legitimate and illegitimate sources of authority'? How does authority cope with transgression? To take one example, the adult western of the middle and late 1950s, with its drama of solitary righteousness and suppressed libidinousness, can be seen in retrospect to have played on a subterranean casualness about authority that was at odds with the dominant hierarchical motif of the Eisenhower years: TV drama here was vaguely premonitory of a counterculture that had not yet crystallized into social action.

Richard Boone's Paladin in "Have Gun, Will Travel" and James Garner's Bart Maverick in the series of the same name were lone heroes standing solidly within the tradition of frontier insouciance. Like classical good-guy western moralists, they took official law-and-order wryly." And yet, unlike the Lone Ranger and his puritanical ilk, they mixed their pursuit of outlaws with a pursuit of pleasure; they were hedonists. In the meantime, Matt Dillon of "Gunsmoke" was a decent Eisenhower-like public official, affirming the necessity of paternalistic law and order against the temptations of worldly pleasure (Kitty, the saloonkeeper? madam?) and the depredations of unaccountably wicked outlaws. in the early 1960s the western declined, and with the rise of Camelot counterinsurgency and the vigorous "long twilight struggle" of John F. Kennedy, its values were taken over into spy-adventure drama. Series like "Mission: Impossible" and "The Man from UNCLE" capitalized on the CIA's mystique and reined individualism into teamwork; such shows were more or less synchronized with official government policy. "Star Trek," launched in 1966, continued the teamwork and processed the cast's internationalism (intergalacticism, rather, if we allow for Spock) into a benign interstellar imperialism: the disordered universe, full of misguided utopians and deceitful aliens, needed a continuing United Nations police action operating under the cool white American head of Captain Kirk.

Police shows also display a metamorphosis that matches the decomposition of the dominant view of crime and punishment. In Jack Webb's "Dragnet," beginning in 1952, the police are in harmony with society's values; they detect according to the book; they are pure technicians ("Just the facts, ma'am"), representing the coincidence of technical and political capacities in the state. Organization is strictly hierarchical: Sergeant Friday's authority is undisputed. Crime never pays; each installment ended by telling the audience that the criminal was convicted and sentenced to a definite term in prison. But by the late 1960s, the social consensus about the decency and the effectiveness of the state has unraveled, and the next generation of police shows displays uncertainty about the legitimacy and consequence of the law, and of the police within it, and about the organization of authority within the police. Into the 1980s, one continuing message is the practical futility of liberalism, a sense imported from the larger political culture.

In cop and detective shows, there are a variety of hybrid mixtures of authority and outlawry, elaborating, in turn, a range of popular ambivalences toward bureaucracy, law, and the state. Hierarchy is no longer taken for granted and is no longer harmonized with effective law enforcement. Insider official authority and outsider cowboy integrity and restlessness have been combined and condensed into the character of the private detective, halfway between the police force (which continually gets in his way and must be outfoxed) and the criminal (who shares his resistance to corporate norms, yet testifies to the permanence of evil).

The ex-cop private detective, from fiction's Lew Archer to TV's Harry 0, is the anarchist as refugee from the organization. He shares the goal of the police but disdains the standard rules. He is the classic American frontier individualist, but in the service of a law and order whose primary institutional embodiment, the police force, he scorns. He is half anarchist, half vigilante. He represents the individualist's partial resistance, partial accommodation to a bureaucratic order that conditions his own ideals and yet cannot contain his spirit. Through his persona, scriptwriters who are confined to the organized strictures, however remunerative, of series formulae pay tribute to the glimmering image of the autonomous writer they want to be.

This imago, straining at the social leash, speaks to the frustrated aspirations of employees in living rooms- So, in a different way, does Kojak, the antibureaucratic cop trapped in red tape, scornful of criminal-coddling officials who are pushed around by the courts and let legal niceties stand in the way of rough justice. In still different ways, so do the various disillusioned officers of "Hill Street Blues" and the corner-cutting heroes of "Miami Vice."

The transformations in other genres also register shifts and variations in the condition of hegemonic ideology. The technologically enhanced superhero, for example, has metamorphosed over the course of four decades, as Thomas Andrae's study of Superman delineates. In the 1960s, the straight-arrow Superman was supplemented by the whimsical, self-parodying Batman and the Marvel Comics heroes, symbols of power gone slightly silly, no longer prepossessing. In playing against the conventions, their producers were doubtless affected by the modernist self-consciousness so popular in high culture at the time. Thus do shifts in genre presuppose the changing mentality of critical masses of writers and cultural entrepreneurs; yet these changes would not take root commercially without corresponding changes in the disposition (even the self-consciousness) of large audiences. Changes in cultural ideals and in audience sensibilities must be harmonized to make for shifts in genre or formula.

If I have left the impression that television entertainment ordinarily tilts toward liberal variants of the hegemonic worldview, this is an apt moment to correct the picture. Although at this writing technological superheroes are missing from prime-time television, their most recent incarnations, in the 1970s, corresponded to militarist tendencies in the American polity. The Six Million Dollar Man and the Bionic Woman were not only obediently patriotic, they were organizational products through and through. Such team players had no private lives from which they were recruited task by task, as in "Mission: Impossible"; they were not fortuitous arrivals from another planet, nurtured by sturdy farm folk, like Superman; they were rescued by the state and equipped by it. They owed their very existence to the state's fusion of moral right and technological know-how. And occasional topical slants anchored these shows' general support of military solutions to international problems. One 1977 episode of "The Six-Million-Dollar Man," for example, told the story of a Russian-East German plot to stop the testing of the Air Force's new B-1 bomber; by implication, it linked the domestic movement against the B-1 to the foreign red menace.

Setting and Character Types

Just as prevailing television genres shift in historical time, so do the settings and character types associated with them. Shifting market tolerances and producer interests make for noticeable changes, some of which we have already discussed. Even in the formulaic 1950s, a few comedies were able to represent discrepant settings, permitting viewers both to identify and to indulge their sense of superiority through comic distance. Jackie Gleason's "The Honeymooners" and "The Phil Silvers Show" (Sergeant Bilko) capitalized on their stars' enormous personal popularity and theatrical experience, and were able to confer dignity on working-class and deviant characters in situations the opposite of glamorous.

Indeed, these examples point to the general importance of stars as such in binding the audience to the show. Stars who exude sexual auras are, of course, especially alluring, although talent scouts are notoriously unable to predict which auras are going to work. The audience is in part, then, psychologically involved with the actor, who is perceived beneath the momentary mask of the characte. But all television (like film) rests on libidinal identifications with characters, whether explicitly sexual or not. Audiences invest stars with powers that descend from residues of infantile experience, experience that comes back into play during the provisional regressions of viewing. It is a nice fantasy that particular stars embody particular aspects of hegemonic ideology, particular dimensions of the audience's structures of feeling, most likely in transposed forms; but how this embodiment takes place concretely, in particular stars at particular times, requires a full-blown analysis of its own.

Suffice it to say, for the moment, that two uniformities of setting and character underlie almost all the variation. For one, the set itself almost always propounds a vision of consumer happiness. Living rooms and kitchens usually display what David Riesman has called the standard package of consumer goods. Even where the set is ratty, as in "Sanford and Son," or working class, as in "All in the Family," the bright color of the TV tube almost always glamorizes the surroundings so that there will be no sharp break between the glorious color of the program and the glorious color of the commercial. In the more primitive 1950s, by contrast, it was still possible for a series like "The Honeymooners" or "The Phil Silvers Show" to get by with one or two simple sets per show: the life of a good skit was in its accomplished acting. But that series, in its sympathetic treatment of working-class mores, was exceptional. Color broadcasting accomplishes the glamorous ideal willy-nilly.

Second, the major characters are winners. Most of the time, NETWORKs and sponsors want to convey images of glamor and fun; the settings and characters must be showcases not just for commercials but for an entire fantasy world that will entice the mass audience. Although program control has shifted from sponsors to NETWORKs-with certain implications for settings and character types, as we shall see below-what has not changed is television's preference for winners, In 1954, for example, one advertising agency wrote to the playwright Elmer Rice explaining why his Street Scene, with its "lower class social level," would be unsuitable for telecasting:

"We know of no advertiser or advertising agency of any importance in this country who would knowingly allow the products which he is trying to advertise to the public to become associated with the squalor... and general "down" character... of Street Scene.

On the contrary it is the general policy of advertisers to glamorize their products, the people who buy them, and the whole American social and economic scene.... The American consuming public as presented by the advertising industry today is middle class, not lower class; happy in general, not miserable and frustrated."

Twenty-five years later, it is the NETWORKs that are directly enforcing this cheer. Television's professionals must get results; they must return each week, like fixed stars, to embody the upbeat." Bob Shanks, an ABC vice-president who has written a revealing insider's book about television, explains bluntly why he would advise against a series about a black ex-convict: "Perry Mason must win every week. So must Dr. Welby." Shanks's rationale reveals the mentality of television executives, if not necessarily the whole audience: "Comedians and social critics may scoff; we ourselves know life is not like that. So what? People, masses of people, do not watch television to learn what life is like, but rather to escape it. Defeat and dreariness are what happen to you during the day. At night, in front of the box, most people want to share in victories, associate with winners, be transferred from reality."

For dramatic purposes, victory should be hard won. There is nothing more boring than the inevitable. Therefore, week after week, the hero should confront forces that are convincingly wicked, whether social (the cop's "bad elements") or natural (the doctor's diseases). In either case, wickedness usually erupts outside social contexts; it has no deep cause. It happens, it needs to be fixed, period. The melodramatic need for the service justifies the hero's power in the situation. Clients, fools that they are, often start out recalcitrant, then turn out cooperative as they discover what is good for them. The patient is part of that intractable material world that makes the professional's job so difficult-yet, in the end, so rewarding, so necessary, so worthy of the prestige that attends it.

And yet knowledge is not the decisive attribute of the hero's status; he (or, rarely, she) deserves respect by virtue of his personality. Dr. Marcus Welby was not just any doctor; he was the persona created by the paternal Robert Young we already knew best for his crinkly smile and moral excellence. Even if gruff, he was warmhearted, dedicated to keeping families together. "Father Knows Best"... Welby Knows Best. Many of the most successful shows were adept at fusing skill and character in the same hero. In the highest form of this fusion the actor succeeds in investing himself with the character's prowess, carries his aura from character to character, and also exploits it commercially.

Karl Malden (playing his "Streets of San Francisco" character) advertises American Express Travelers Checks to foil crooks; Robert Young heaps praise on decaffeinated coffee to ease tension. As I have pointed out above, it is in the nature of the viewing process that some portion of the audience-clientele will respond to this blurring of fact and fiction: during the years that "Marcus Welby" enjoyed such popularity, Robert Young got five thousand letters a week asking for medical advice.

The senior professional hero is also a man fixed in the present. Whatever he knows, he always knew. He did not arrive; he was not recruited; he was not trained; he did not come from this or that class; he did not go to these or those schools. We know him almost entirely by his works and his personality." Except for an occasional origin myth, which tells a story about how the man got where he is, he is never a man in formation. Thus his prowess is presented as something magical. Only his understudy, the stereotypic up-and-coming young acolyte, can hope to learn what the older man knows. As for clients, they are lucky to be flattered with his attention. But lower-status professionals, though they too have come out of nowhere, are less competent. According to Michael Real many real-life nurses wrote letters to the producer of "The Nurses" "complaining of the ineptness of the student nurse in the program, failure to show her in a student role, overly dramatic presentations of hospital life and the nurse's role, and portrayals of nurses as alcoholics, reactionaries, and neurotics." Above all, the televised professional is an idealist and an individualist. His motives are pure: to restore a warm, decent status quo ante, the symptom-free family or the crime-free neighborhood. The fee is never much of an issue, if it is mentioned at all. Perry Mason accepted a retainer with a quick flat grin, the impersonal gesture of a cash register, before getting on with the job he was born to do. Doctors have been even more exempt from the marketplace. As David W. Rintels has said, "Anyone who watches Marcus Welby, M.D., Medical Center, and The Doctors... must of necessity believe... no doctor ever charges for his services; no hospital ever bills a patient; no one ever has to go on charity, or do without care." A good deal of the reason is that the great majority of the patients can afford to pay their own way handsomely; Michael Real has shown, for example, that almost all Welby's patients were well-to-do, Cops may occasionally grumble about their salaries: this is in keeping with the more realistic tone of the cop show. But, cops and teachers aside, the televised professional is generally a free professional.

We have, then, what seems at first glance distortion pure and simple. While the actual professions have become more bureaucratized and specialized, the television hero upholds the traditional image of the self-sufficient practitioner. The doctor has been either an omnicompetent general practitioner or a surgeon, that most prestigious of medical specialists, embodying a power of life and death that lends itself easily to melodrama. In either case, he does not condescend to bureaucratic relations with Medicaid or Medicare or Blue Shield or the hospital administration; he is not hedged about by insurance companies; he does not politic through professional organizations because he does not need to, His hands are not dirtied by the complicated structures of his everyday world.

He is free to be a household god: he is a romantic figure. Even the cop, forced to work within the confines of bureaucracy, as we have seen, is the romantic as organization man, his rebellion always incipient, always stylistic, always idealistic, always solitary. In the real America, the professional carries forward an older tradition: the frontiersman, the sawbones, the Lone Ranger. Almost everyone in the audience works for someone else; only about 2 percent of the working population are unsalaried professionals. But television conventions allow the audience its free-standing heroes: repositories of freedom and nurturance united, and dreams of independence for the kids.

For if television is untrue to the reality of society, it is true to a dream; it pays tribute to popular fantasy. The televised image both anchors and reinforces the prevailing aspiration toward professional status. In 1962, less than 2 percent of the sons of manual laborers had entered the professions; but the aspiration persists, the luminous hope that the child will be a successful and respectable professional, will be able, as Richard Sennett writes, "to unite love and power." So it is fair to criticize television images for lacking naturalistic accuracy, for racist and sexist and class-biased stereotyping; yet in one crucial respect the criticism misses the point. Television, like much popular culture through the ages, embodies fantasy images that speak to real aspirations. It does not simply reflect the social world; it is no mirror. The hegemonic image is an active shaping of what actually exists, but it would not take hold if it did not correspond, one way or another, to strong popular desires-as well as to defenses against them. "False consciousness" always contains its truth: the truth of wish, the truth of illusion that is embraced with a quiet passion made possible, even necessary, by actual frustration and subordination.

But to say that the hero corresponds to a popular wish is not to say that the wish constructs the image. It is the culture industry that generates the image, often enough tailoring it to the public relations desires of the actual profession. The NETWORKs guarantee that the professional image will not be too starkly, consistently, "controversially" idealistic by setting up direct and indirect systems of censorship. After self-censorship has had its chance, most of the work of censorship is done in-house, by the NETWORKs' own censorship bureaus. Earlier, some of this work was farmed out to professional associations such as the American Medical Association, which had first crack at approving the televised images of their professions.

We have already noted the case of the FBI. And as for medicine, we have the testimony of Norman Felton, the executive producer of "Dr. Kildare," "The Eleventh Hour," and "The Psychiatrist," that "on the Dr. Kildare series we were asked by NBC to get the approval and seal of the AMA. This meant that we submitted scripts for approval to the AMA. Although the organization gave us technical help, it goes without saying that we did not present an accurate picture of the practice of medicine, or the difficulties many people had in obtaining medical care."

But in the end it is the NETWORKs that confer censorship power. They operate within a web of hegemonic institutions; their oligopoly is interlinked with the prestigious monopolies, where power and knowledge interpenetrate; and the NETWORKs act in behalf of their conception of the whole. They may well relax censorship over dirty words and sexual innuendo in order to keep up with the changing standards of their younger, hipper audiences (and of Hollywood); but they work to secure a complexly articulated version of legitimate social authority—which needs to be traced out in detail. The NBC Radio and Television Broadcast Standards and Practices manual reads: "Respect for lawyers, police, teachers and clergy should not be diminished by undue and unnecessary emphasis on unfavorable aspects of members of these professions". Bob Shanks of ABC writes that "all three NETWORKs have similar policy guidelines and adhere to the same industry and government codes."

Again, although most hegemonic constancies remain, shifting market structures make variation possible. The near universality of television set ownership creates the possibility of a wider range of audiences than existed in the 1950s. Minority-group, working-class, age-segmented, and subculturally compact audiences have proliferated. Since more than half of American households own two or more television sets, it becomes possible to target relatively narrow bands of the population. Programming becomes more centrifugal. Cable and pay television multiply, and the market becomes more segmented. The glamor standards can slacken off at times. The industry noted that the miniseries "Roots" reached people who don't normally watch TV, with the homes-using-television levels during the week "Roots" was aired up from 6 to 12 percent over the comparable week a year earlier.

Untapped markets can only be brought in by unusual sorts of programming. There is room in the schedule for rebellious human slaves, just as there is room for hard-hitting technological superheroes. Movies made for single showings on television also may veer toward counter hegemonic political positions: they may sympathize with homosexuals, or they may criticize Senator Joseph McCarthy and the blacklist of the 1950s. ABC's 1980 version of Attica, based on Tom Wicker's A Time to Die,focused its first half on the grievances of prisoners and construed subsequent events in that light; ABC did insist that the producer splice in new footage at the last moment, showing prisoners' knives at the throats of their hostage guards, but the power of the show's reformist sympathy for the prisoners remained powerful. Such shows have prestige value and help deflect critical opinion from the NETWORKs.

NETWORK elites do not, however, invest in regular heroes who will challenge, from left, right, or elsewhere, the core values of corporate capitalist society-who are, say, union organizers or explicit socialists or, for that matter, born-again evangelists. And the emergence of independently syndicated programs for independent stations does not necessarily translate into increased diversity of substance. The everyday settings and character types of television entertainment go on confirming the essential soundness of commercial values, the prerogatives of individualism, the authority of professionals (though not political leaders or business executives), and the legitimacy of the national security state.

Conflict and Solution

All narratives set out problems and point to solutions. High art works its problems through with absorbing thoroughness, toward genuine resolutions; much commercial art, slapped together under pressure by formula, fails to justify its own feeble, jerry-built passes at solutions. But solutions, in either case, there must be. For it is in solution that the tensions provoked in the audience by the shaping and disguising of their desires and fears get resolved; it is through solution that the work finally produces its psychological effect. Through happy ending, that harmonious result which reality denies may be granted by fantasy; the irreconcilable may be reconciled But to say that solution is important is not to say how a given cultural form will frame its solutions. There must be conventions; but which will prevail? The essence of the dominant television convention is that, whatever social and psychological problems come up, they are susceptible to successful individual resolutions. However grave the problems, however rich the imbroglio, the episodes regularly end with the click of closure: an arrest, a defiant smile (now held with freeze-frame technology), an 1-told-you-so explanation before the credits. As we have already seen in the discussion of character types, the convention of the starbased series requires that preserving solutions be found; it would hardly do for the hero to fail week after week, or to be killed in the line of duty.

The characters with which the audience identifies must stay alive and well, ready for next week's imbroglio. (Only when there is a contract dispute or an actor departs for more alluring horizons or dies, does a character get written out of the script. The more sympathetic the character, the more gentle the departure.) However deeply the problem may be located within society, it will be solved among a few persons: the heroes must attain a solution that leaves the rest of society untouched. Crime is solved by arrests; citizens never organize against it. Meanness is resolved by changes of heart or the bad guy's " come-uppance. "

Again, there are variations alongside the normal convention. There was the short-lived "East Side, West Side," and Norman Lear's independently syndicated "Mary Hartman, Mary Hartman," which extended soap opera conventions in the process of parodying them. (The NETWORKs would not buy it.) Beginning in 1978, CBS's successful "Dallas" signaled that the convention of daytime serial continuity-the rolling plot in which one damn thing leads to another-had become acceptable in prime time. Earlier, Norman Lear's archetypal "All in the Family" was unusual among NETWORK broadcasts in sometimes ending obliquely, softly, or ironically, on the curvature of a question mark, thus acknowledging that the Bunkers could not solve a problem whose genesis was outside their household.

"Lou Grant" was even more unusual: for a time it specialized in open rather than closed endings. One show, for example, revolves around Lou's encounter with several young black men who hang out on the street corner. Learning that one outspokenly cynical one is unemployed, Lou, the humane liberal, urges him to apply for a job at the Tribune. The young man, by the end of the show, has failed to get a job, and when Lou returns to the street corner and asks about him, his buddies report that he's left town. His story is still open. The problem remains; individual action cannot eliminate unemployment. According to one of the show's writers, the censors of CBS Standards and Practices, who had not objected to any of the antiestablishment politics implicit in many "Lou Grant" shows, had objected to the show's open form and were trying to regear it to orthodoxy. An even more startling variation from type was ABC's 1983 "The Day After," in which personal action was utterly unavailing. Form as well as manifest content-to the extent they can be distinguished are matters of concern for the television elite. "Roots," for example, represented both the cruelty of slavery and the possibilities of resistance; it humanized blacks and romanticized the possibility of a class alliance between slaves and poor whites, Yet the series also pointed to the chance for upward mobility; the upshot of travail was freedom. Where Alex Haley's book was subtitled "The Saga of an American Family," ABC's version carried the label-and the patriotic and institutional self-congratulation- "The Triumph of an American Family."

Who could say categorically whether the prevailing impression was that of the collective agony or that of the family's triumph? Both themes were there, to be taken seriously in different proportions by different audiences. Those elements conflicted within a complex whole and helped renew the hegemonic understanding of right relations among nation, family, and individual. The friction between agony and triumph reproduced a larger friction. Cultural bargains of this sort keep the hegemonic ideology in motion and in equilibrium at the same time.

9

The Advancement of TV Presentation System

INTRODUCTION

Much analysis of new media is based upon an assessment of competing technologies, investor activities, marketing deals, revenue projections, regulatory issues and price points. All of these factors are important. However, there are other crucial factors that consistently receive less attention in studies about new media, such as content, service design (e.g., user interfaces) and how people actually use a new medium. This discussion will focus on user behavior and the interactions among behavior, technology and programming. It leaves business and regulatory issues for others to assess. When a technology is introduced, it often leads to changes in user behavior over time (e.g., the remote control led to more channel changing). In turn, behavioral changes can lead to new forms of programming that respond to the changed behavior (e.g., channel changing behavior has led to changes in the design and placement of commercials, greater segmentation of content within programs and the placement of icons in the corner of screens to 'flag' the channel name for high speed channel surfers). Programming and behavior changes may then create an opportunity for a next generation of technology that takes advantage of the new environment. For example, personal video recorders (PVRs) are intended to support current television consumption patterns by making the full 24-hour schedule of large capacity cable and satellite services available to viewers with busy lifestyles and limited time to watch TV. The cycle of interactions among technology, behavior and programs can continue over many years or decades.

There is a second missing piece in much analysis of new media - an historical perspective. In the rush to predict where we are heading with new technologies, we often ignore the past, assuming that it has little to teach us about the future. For example, what could the introduction of television in the 1930s and 1940s possibly tell us about the introduction of interactive television in 2002? This discussion takes a different tack and argues that historical patterns are very useful in guiding assessments of new media. For example, interactive television was introduced in a number of trials and services during the 1990s without much success. Did this marketplace experience signal the absence of consumer interest in ITV or was it a false start? History demonstrates that television itself as well as VCRs and videodiscs, among many other technologies, experienced false starts. They were introduced and failed or experienced limited success, only to be re-introduced and succeed at a later point in time. The reasons why they failed initially and subsequently succeeded provide useful lessons for ITV.

This discussion focuses on television viewing behavior beginning in the late 1930s and traces changes in the use of television through our current environment, then speculates about some changes to come. It has an admitted North American bias, owing to the author's access to archival materials on the western side of the Atlantic and my ethnographic research work in U.S. households. However, some European examples will be cited. Further, the audience is invited to correct this bias and provide a more complete picture of TV viewing behavior.

TELEVISION THROUGH THE DECADES AND THE WAYS IT CHANGED OUR WORLD

Arguably, the greatest and most influential invention of the 20th century, perhaps with the exception of the Internet, the television began with humble beginnings and was met with skepticism. However, since its inception, almost 75 years ago, the United States has been captivated. Now, according to USA Today and Nielson Media Research, the average American home has 2.71 televisions and 2.55 people. There are more TVs in the average home than there are people. At the turn of the century, the TV as we know it today was only a dream in the eyes of inventors. Many believed they could take the basic concepts applied to the transmission of sound and apply them to the transmission of pictures. It would be unfair to say the invention of the television is attributed to one person. Philo T. Farnsworth, Vladimir Zworykin, Charles Jenkins and John Baird all attributed greatly to the invention of the modern television each supplying their own piece of the puzzle. Zworykin especially, when he used a small piece of technology called cathode ray tube (CRT) in the receiver. Although seen as primitive compared to modern television advancements like Plasma and LCD, the CRT is still used in TV sets today. Since its inception in the early 1930's and commercial availability in the late 1930's the evolution of the television is interesting. Although it seems like a world away, it has only been in recent years that the technology has changed drastically. Here is a time line to demonstrate the look, feel and uses of the TV over the decades.

1930s

After its formal introduction at the 1939 World's Fair, the first real commercial televisions became widely available. However, many networks like NBC and CBS had already been broadcasting for several years. Radio networks were quickly adjusting their studios to accommodate the new medium. Sets were large pieces of equipment with about 12 inch screens. They cost about $400 to $500 and the average household income was about $1300 a year. Programming was sparse. There was no "must see TV," networks broadcasted specials like the 1932 presidential election, boxing matches and news reports. One of the most recognized images from this early era includes the 1936 Olympics from Berlin in which Hitler announced the opening of the ceremonies.

1940s

Much of the hype surrounding televisions in the 1940's was swept under the rug with the onset of WWII. Although in the earlier portion of the decade, the United States wasn't involved with the war, production of televisions stopped so the U.S. could put their efforts toward developing radar technology. Even though the production of televisions stopped, the innovation persisted and color was introduced at the early part of the decade. This is also when the television commercial was invented.

During the war, the television was used as a small propaganda machine. Encouraging people to buy bonds and support the effort. The post-war economy was strong and the late 40's saw a boom in television production. Two types of models surface the tabletop and console. People like Milton Berle, Ed Sullivan and Howdy Doody become some of TV's first stars.

1950s

Often the 1950's have been seen as the hay day of the economical boom in the United States and the Golden Age of television.

More people are buying consumer electronics, by the end of 1951 and there are more than 8,000,000 TVs in the United States.

I Love Lucy, The Honeymooners, Father Knows Best and the Lone Ranger are some of the most watched programs. The TV Guide is the #1 magazine in the country. Color becomes the wave of the future and the Remote Control is invented. Although most people had neither.

The "tube" becomes an advertiser's dream come true. Products are marketed directly to the consumer by sponsorships. The "soap opera" is thus named for the cleaning products that sponsored the daytime dramas targeted directly at women. The family dinner is revolutionized with the introduction of the TV dinner. Family meals will never be the same. By the end of the decade, a 21-inch black and white set was about $200 and the average 21-inch color set was almost $500.

1960s

Often the 60's are seen as the end of an era. An age of innocence seen in the 1950s died and would never be seen again in the United States. After the assignation of President Kennedy, families gathered around to watch the news depicting a changing social climate and growing hostility toward an oppressive government. Americans begin to get more news from the television than the newspaper. The United States became obsessed with space travel. Everything sold had a space age edge, especially televisions. Everyone gathered to watch shuttle launches and moon landings. Televisions became more portable and affordable. It is estimated 75 million people watched the Beatles on the Ed Sullivan Show. The Vietnam War is the first conflict to be televised. Doctors legally advertised cigarettes. By the end of the decade there were approximately 78 million television sets in homes across the United States; 200 million around the world.

1970s

Sesame Street debuts to children across the country. The show is still on the air after almost 40 years. Nighttime broadcasts broke the traditional molds. Couples were now sleeping in the same beds, actresses could show their belly buttons and "All in the Family" and "MASH" are two of the most popular evening programs. Although it wasn't the first home video game, Pong became the most popular. The average salary is 7,500 a year and the average price for a TV was between $400 and $700 dollars. The first direct to broadcast satellite television was launched in 1972. 1978 was the last year large Black and White consoles were manufactured.

1980s

The 80's saw little innovation with the television specifically; however, there became a growing number of television accessories like the VCR and home game consoles like Nintendo. Although cable had been around since the 50's, cable television saw a significant boom during the 80's.

Sitcoms were more popular than ever. The 80's spawned programs like "Rosanne," "The Cosby Show" and "Married with Children" (which is the longest running sitcom in television history.) The VCR took the United States by storm. Although introduced in the late 70's, the VCR didn't become a staple in the American home until the mid 1980s. Now, people could watch movies in the comfort of their own home and record and watch their favorite programs on their own timetable.

1990s

The 90's saw big changes in technology all together. One can arguably say the 1990s was an age of technological change, thanks largely to the Internet. The personal computer became affordable enough for average people to own, the Internet was available through dial-up ISPs like Prodigy and AOL, and information was seemingly at our fingertips.

Television programming became more risqué, pushing the FCC's buttons and spawned some of the most loved shows of all time including "Friends," "Seinfeld," "Twin Peaks" and "The Simpsons."

Johnny Carson left the Tonight Show in 1993 after more than 30 years on the air. Cable television programming saw a large boom at the end of the decade with shows like "The Sopranos" and "Sex in the City." Broadcast on HBO, these series were exempt from standard FCC regulations and frequently depicted scenes with graphic violence, gratuitous sex and foul language. Audiences and critics loved it.

Various TV technologies like Plasma and LCD are in experimental stages. Standard CRT televisions still lead the market. However, televisions now include additional features like picture in picture, sleep timers and parental controls.

2000s

DVD players take over the home theater experience. Much like the VCR, DVD players were introduced a decade earlier, but took some time to gain momentum. At the beginning of the decade, DVD players were in approximately seven percent of homes; in less than 10 years, more than 80 percent of homes had a DVD player.

Thin is in. Advances in LCD and Plasma technology enable television manufacturers to produce a better picture, larger screen sizes and save space.

The "home theater" experience was in the forefront of manufacture's minds providing bigger and better quality to the consumer. TiVo revolutionized the way we watch television. TiVo had the ability to pause live broadcasts and record several programs at once, even while the television was on another channel. TiVo can schedule recordings based on time, channel, title or celebrity. No longer were people tethered to their couches living by the network's times. They were free to watch what they wanted when they wanted.

Reality shows take over the airwaves and TV becomes interactive. With the introduction of "American Idol" home viewing audiences became part of the competition. Encouraging the public to "vote" for their favorite performer, the programs enabled audiences to take an active part in the show's outcome. Now, a 30-second ad during "American Idol" costs about $750,000. Television goes all digital. In 2009, all analog televisions will go black. Networks will only be producing a digital signal.

Less than 10 years after the DVD player's introduction, the second wave of video players were introduced touting better quality and longer run times. HD DVD and Blu-ray enter a battle for high definition supremacy, which is reminiscent of the VCR and Betamax war 30 years prior.

TV goes online. With the introduction of high speed Internet, video streaming and digital recording enter the scene. Many are posting their favorite programs and commercials online without the permission of the broadcasters. This has led to many networks making their programs available online. Seventy-five years ago, the television was introduced with skepticism and awe. No one truly believed it would change the way we view the world. Now, people are more attached to their televisions than ever including programming on computers and cell phones; and manufactures, broadcasters and producers are continually finding new ways to bring big entertainment to the small screen.

EARLY TV VIEWING BEHAVIOR

The earliest photographs of TV watching show people sitting relatively close to TV screens - one to four feet away. This was associated with the small size of the screens (typically, 10 to 14 inches in the 1930s through the early 1950s) and fuzzy images.

Typically, there was only one TV per household and a very limited number of channels. There was little channel changing and TV viewing was often in groups. In the late 1940s and early 1950s, the group consisted of family members and, commonly, neighbors who did not yet have a television. People invited neighbors over to their house to watch TV so that they could enjoy the new medium and also to show off this expensive status symbol. In 1947, the price of a TV in the U.S. was six weeks salary for an average household. This was a relative bargain compared to 1939, when TVs cost nearly as much as a car.

Early photographs of TV viewing also show that there was a lot of TV watching in public places such as department stores, bars and the windows of many different types of stores that simply wanted to draw people near the store. This is supported by some interesting data. In 1947, three million people watched the final game of the World Series on television and 90 percent of those people watched in a public location such as a department store or a bar. Bars in particular used television to draw in customers who did not have a TV at home. Public TV watching also built an appetite for the new medium and millions of people saved up over many months in order to purchase a TV for their home.

Many TVs in the 1930s and 1940s were contained within fancy pieces of furniture. Perhaps this helped to soften the blow of the enormous price. By the 1950s, many TVs were manufactured in plain metal shells, much like TVs today. This too may have been related to pricing strategies and attempts to bring down the cost of television sets for the mass public that was beginning to buy them. During this time, another curious sociological phenomenon emerged. Photos from this time period show that some families began to place photographs of family members on top of or next to TV sets. This probably reflects the increasing time people were spending with TV and the acceptance of television as "part of the family." By the late 1950s, and the introduction of color TV, many TV sets were once again encased within furniture. This may have been a repetition of an earlier pattern - early color technology drove the price of TV sets back up to six weeks salary for an average household. Wooden cabinets may have helped to justify the high price. Among the more lavish furniture designs was a German television set, the "Kuba Komet" in 1960. It was housed in a blond wooden shell with sharp angles and served as a full entertainment center, with a phonograph and radio as well as a TV screen.

THE BEGINNING OF COMMERCIAL TELEVISION

By 1949 Americans who lived within range of the growing number of television stations in the country could watch, for example, *The Texaco Star Theater* (1948), starring Milton Berle, or the children's program, *Howdy Doody* (1947Ð60). They could also choose between two 15-minute newscastsÑ*CBS TV News* (1948) with Douglas Edwards and NBC's *Camel News Caravan* (1948) with John Cameron Swayze (who was required by the tobacco company sponsor to have a burning cigarette always visible when he was on camera). Many early programsÑsuch as *Amos 'n' Andy* (1951) or *The Jack Benny Show* (1950Ð65)Ñwere borrowed from early television's older, more established Big Brother: network radio. Most of the formats of the new programsÑnewscasts, situation comedies, variety shows, and dramasÑwere borrowed from radio, too (see radio broadcasting and television programming). NBC and CBS took the funds needed to establish this new medium from their radio profits. However, television networks soon would be making substantial profits of their own, and network radio would all but disappear, except as a carrier of hourly newscasts. Ideas on what to do with the element television added to radio, the visuals, sometimes seemed in short supply. On news programs, in particular, the temptation was to fill the screen with "talking heads," newscasters simply reading the news, as they might have for radio. For shots of news events, the networks relied initially on the newsreel companies, whose work had been shown previously in movie studios. The number of television sets in use rose from 6,000 in 1946 to some 12 million by 1951. No new invention entered American homes faster than black and white television sets; by 1955 half of all U.S. homes had one.

McCARTHYISM

In 1947 the House Committee on Un-American Activities began an investigation of the film industry, and Sen. Joseph R. McCarthy soon began to inveigh against what he claimed was Communist infiltration of the government. Broadcasting, too, felt the impact of this growing national witch-hunt. Three former members of the Federal Bureau of Investigation (FBI) published "Counterattack: The Newsletter of Facts on Communism," and in 1950 a pamphlet, "Red Channels," listed the supposedly Communist associations of 151 performing artists. Anti-Communist vigilantes applied pressure to advertisersÑthe source of network profits. Political beliefs suddenly became grounds for getting fired. Most of the producers, writers, and actors who were accused of having had left-wing leanings found themselves blacklisted, unable to get work. CBS even instituted a loyalty oath for its employees. Among the few individuals in television well positioned enough and brave enough to take a stand against McCarthyism was the distinguished former radio reporter Edward R. Murrow. In partnership with the news producer Fred Friendly, Murrow began *See It Now*, a television documentary series, in 1950. On Mar. 9, 1954, Murrow narrated a report on McCarthy, exposing the senator's shoddy tactics. Of McCarthy, Murrow observed, "His mistake has been to confuse dissent with disloyalty." A nervous CBS refused to promote Murrow and Friendly's program. Offered free time by CBS, McCarthy replied on April 6, calling Murrow "the leader and the cleverest of the jackal pack which is always found at the throat of anyone who dares to expose Communist traitors." In this TV appearance, McCarthy proved to be his own worst enemy, and it became apparent that Murrow had helped to break McCarthy's reign of fear. In 1954 the U.S. Senate censured McCarthy, and CBS's "security" office was closed down.

THE GOLDEN AGE

Between 1953 and 1955, television programming began to take some steps away from radio formats. NBC television president Sylvester Weaver devised the "spectacular," a notable example of which was *Peter Pan* (1955), starring

Mary Martin, which attracted 60 million viewers. Weaver also developed the magazine-format programs *Today,* which made its debut in 1952 with Dave Garroway as host (until 1961), and *The Tonight Show,* which began in 1953 hosted by Steve Allen (until 1957). The third network, ABC, turned its first profit with youth-oriented shows such as *Disneyland,* which debuted in 1954 (and has since been broadcast under different names), and *The Mickey Mouse Club* (1955Ð59; see Disney, Walt).

The programming that dominated the two major networks in the mid-1950s borrowed heavily from another medium: theater. NBC and CBS presented such noteworthy, and critically acclaimed, dramatic anthologies as *Kraft Television Theater* (1947), *Studio One* (1948), *Playhouse 90* (1956), and *The U.S. Steel Hour* (1953). Memorable television dramas of the eraÑmost of them broadcast liveÑincluded Paddy Chayefsky's *Marty* (1955), starring Rod Steiger (Ernest Borgnine starred in the film), and Reginald Rose's *Twelve Angry Men* (1954). By the 1955Ð56 television season, 14 of these live-drama anthology series were being broadcast.

This is often looked back on as the "Golden Age" of television. However, by 1960 only one of these series was still on the air. Viewers apparently preferred dramas or comedies that, while perhaps less literary, at least had the virtue of sustaining a familiar set of characters week after week. *I Love Lucy,* the hugely successful situation comedy starring Lucille Ball and Desi Arnaz, had been recorded on film since it debuted in 1951 (lasting until 1957). It had many imitators. *The Honeymooners,* starring Jackie Gleason, was first broadcast, also via film, in 1955 (lasting until 1956 with the original cast). The first videotape recorder was invented by Ampex in 1956 (see video; video recording; video technology). Another format introduced in the mid-1950s was the big-money quiz show. *The $64,000 Question* (1955Ð58) and *Twenty-One*(1956Ð58) quickly shot to the top of the ratings. In 1959, however, the creator of *The $64,000 Question*, Louis C. Cowan, by that time president of CBS television, was forced to resign from the network amid revelations of widespread fixing of game shows (see Van Doren, Charles).

TELEVISION AND POLITICS

Television news first covered the presidential nominating conventions of the two major parties, events then still at the heart of America politics, in 1952. The term "anchorman" was used, probably for the first time, to describe Walter Cronkite's central role in CBS's convention coverage that year. In succeeding decades these conventions would become so concerned with looking good on television that they would lose their spontaneity and eventually their news value. The power of television news increased with the arrival of the popular newscast, *The Huntley-Brinkley Report,* on NBC in 1956 (see Huntley, Chet, and Brinkley, David). The networks had begun producing their own news film. Increasingly, they began to compete with newspapers as the country's primary source of news (see journalism). The election of a young and vital president in 1960, John F. Kennedy, seemed to provide evidence of how profoundly television would change politics. Commentators pointed to the first televised debate that fall between Kennedy, the Democratic candidate for president, and Vice-President Richard M. Nixon, the Republican's nominee.

A survey of those who listened to the debate on radio indicated that Nixon had won; however, those who watched on television, and were able to contrast Nixon's poor posture and poorly shaven face with Kennedy's poise and grace, were more likely to think Kennedy had won the debate. Television's coverage of the assassination of President Kennedy on Nov. 22, 1963, and of the events that followed, provided further evidence of the medium's power. Most Americans joined in watching coverage of the shocking and tragic events, not as crowds in the streets, but from their own living rooms. A newscast that would soon surpass the popularity of Huntley-Brinkley, *The CBS Evening News with Walter Cronkite,* debuted in 1962 (and was broadcast until 1981). By the end of the decade Cronkite had become not just a highly respected journalist but, according to public opinion surveys, "the most trusted man in America." His role in coverage of the Vietnam War would be important. While the overwhelming majority of television news reports on the Vietnam War were supportive of U.S. policy, television news film of the fighting sometimes gave Americans back home an unfamiliar, harsh, and unromantic view of combat. Many believed it contributed to growing public dissatisfaction with the war. And some of the anger of those defending U.S. policy in Vietnam was leveled against television news.

In 1965, CBS reporter Morley Safer accompanied a group of U.S. Marines on a "search and destroy" mission to a complex of hamlets called Cam Ne. The Marines faced no enemy resistance, yet they held cigarette lighters to the thatched roofs and proceeded to "waste" Cam Ne. After much debate, Safer's filmed report on the incident was shown on CBS. Early the next morning the president of CBS received an angry phone call from the president of the United States, Lyndon B. Johnson, accusing the network of a lack of patriotism. During the Tet offensive in 1968, Cronkite went to Vietnam to report a documentary on the state of the war. That documentary, broadcast on Feb. 28, 1968, concluded with what Cronkite has described as "a clearly labeled editorial": "It is increasingly clear to this reporter that the only rational way out will be to negotiate," he said. President Johnson was watching Cronkite's report. According to Bill Moyers, one of his press aides at the time: "The president flipped off the set and said, `If I've lost Cronkite, I've lost middle America.'"

THE THREE NETWORKS AT THE HEIGHT OF THEIR POWER

In 1964 color broadcasting began on prime-time television. The FCC initially approved a CBS color system, then swung in RCA's favor after Sarnoff swamped the marketplace with black-and-white sets compatible with RCA color (the CBS color system was not compatible with black-and-white sets and would have required the purchase of new sets). During the 1960s and 1970s a country increasingly fascinated with television was limited to watching almost exclusively what appeared on the three major networks: CBS, NBC, and ABC. These networks purchased time to broadcast their programs from about 200 affiliates eachÑstations in each of the major cities or metropolitan areas of the United States.

In the larger cities, there might also be a few independent stations (mostly playing reruns of old network shows) and perhaps a fledgling public broadcasting channel. Programming on each of the three networks was designed to grab a mass audience. Network shows therefore catered, as critics put it, to the lowest common denominator. James Aubrey, president of CBS television, doubled the network's profits between 1960 and 1966 by broadcasting simple comedies like *The Beverly Hillbillies* (1962Ð71). In 1961, Newton Minow, then chairman of the FCC, called television a "vast wasteland." Programming became a little more adventurous with the arrival of more realistic situation comedies, beginning with CBS's *All in the Family* in 1971 (broadcast until 1979). Along with situation comediesÑusually a half-hour focused on either a family and their neighbors or a group of co-workersÑthe other main staple of network prime-time programming has been the one-hour drama, featuring the adventures of police, detectives, doctors, lawyers, or, in the early decades of television, cowboys. Daytime television programming consisted primarily of soap operas and quiz shows until the 1980s, when talk shows discussing subjects that were formerly taboo, such as sexuality, became popular.

The three major networks have always been in a continual race for ratings and advertising dollars. CBS and NBC dominated through the mid-1970s, when ABC, traditionally regarded as a poor third, rose to the top of the ratings, largely because of shrewd scheduling.

PUBLIC BROADCASTING

A Carnegie Commission report in 1967 recommended the creation of a fourth, noncommercial, public television network built around the educational nonprofit stations already in operation throughout the United States (see television, noncommercial). Congress created the Public Broadcasting System that year. Unlike commercial networks, which are centered in New York and Los Angeles, PBS's key stations, many of which produce programs that are shown throughout the network, are spread across the country. PBS comprises more than 300 stations, more than any commercial network. Some of the most praised programs on PBS, such as the dramatic series *Upstairs, Downstairs* (1971), have been imports from Britain, which has long had a reputation for producing high-quality television. Perhaps the most influential of PBS's original contributions to American television were the educational program for preschoolers, *Sesame Street*, which first appeared in 1969Ñand is still a popular programÑand a thoughtful news program called *The NewsHour with Jim Lehrer* (1995; originally *The MacNeil/Lehrer NewsHour*, first broadcast in 1975Ñsee MacNeil, Robert and Lehrer, Jim). Among the many special series produced for public broadcasting, *The Civil War* (1990), a five-part historical documentary, was particularly successful and won some of the largest audiences ever achieved by public TV. PBS funds come from three major sources: congressional appropriations (which suffered substantial cuts beginning in 1982), viewer donations, and private corporate underwriters. None of these types of contributions are problem-free. Government funding brings the possibility of government interference. Conservatives, dating back to the Nixon administration, have pressured PBS to make its programming less liberal. The search for viewer donations has led to long on-air fundraising campaigns. And some critics contend that the need to win corporate support discourages programming that might challenge corporate values.

THE RISE OF CABLE

The force that would challenge the dominance of the three major television networks and offer Americans the choice of dozens and potentially hundreds of television channelsÑcable TVÑbegan quietly in a few geographically isolated towns. Large antennas erected in high places gave everyone connected the chance to receive all the channels available in the nearest city. By 1960 the United States had about 640 such CATV (community antenna television) systems. It soon became apparent, however, that the "television deprived" were not the only viewers who might want access to additional channels and additional programming. In New York City, cable operators contracted to broadcast the home games of the local basketball and hockey teams. By 1971 cable had more than 80,000 subscribers in New York. Then networks specifically designed to be distributed by the cable system began to appear: Time Inc.'s Home Box Office (HBO) in 1975; Ted Turner's "superstation," soon renamed WTBS, in 1976; C-SPAN (live broadcasts of the House of Representatives), ESPN (sports), and Nickelodeon (children's programming), all in 1979. Turner followed with the Cable News Network (CNN) the next year.

INTERNATIONAL GROWTH

Television's development followed different patterns in other countries. Often government, not private corporations, owned some, most, or all of the major networks. In Great Britain the British Broadcasting Corporation, the country's dominant radio broadcaster, established and retained dominance over television. The BBC, funded by a tax on the sale of television sets, established a worldwide reputation for producing quality programming. The

Canadian Broadcasting Corporation, also freed by government support from many commercial pressures, was praised by some observers for the seriousness of much of its news and public-affairs programming. France's major television networks were also supported by the government; however, in France that support was seen as encouraging a tilt in news coverage toward the side of whatever party happened to be in power. By the late 1980s and 1990s, as cable and direct-satellite television systems increased the number of channels, the hold of these government-funded networks began to weaken. Most countries around the world began moving more toward the U.S. model of privately owned, advertiser-supported television networks.

POLITICS ADAPTS TO TELEVISION

By the 1980s politicians and government leaders were familiar enough with the workings of television to be able to exploit the medium to their own ends. This seemed particularly apparent during the presidency of Ronald Reagan, himself formerly the host of a television show (*General Electric Theater*, 1954Ð62). Reagan's skilled advisors were masters of the art of arranging flags and releasing balloons to place him in the most attractive settings. They also knew how to craft and release messages to maximize positive coverage on television newscasts. The Persian Gulf War in 1991 provided further proof of the power of television, with pictures of U.S. bombs falling on the Iraqi capital broadcast live in the United States. Both Iraqi and U.S. leaders admitted to monitoring CNN to help keep up with news of the war. However, the U.S. Defense Department, armed with lessons learned in Vietnam, succeeded in keeping most reporters well away from the action and the bloodshed. Instead, pictures were provided to television by the military of "smart" bombs deftly hitting their targets.

NEW TECHNOLOGIES

In the 1980s, home videocassette recorders became widely available. Viewers gained the ability to record and replay programs and, more significantly, to rent and watch movies at times of their own choosing in their own homes. Video games also became popular during this decade, particularly with the young, and the television, formally just the site of passive entertainment, became an intricate, moving, computerized game board. The number of cable networks grew throughout the 1980s and then exploded in the 1990s as improved cable technology and direct-broadcast satellite television multiplied the channels available to viewers.

The number of broadcast networks increased also, with the success of the Fox network and then the arrival of the UPN and WB networks. The share the broadcast networks attracted continued to erode, from well over 90 percent in the early 1980s to under 50 percent by 1997. Although the population of the United States has continued to grow, the Nielson Media Research company estimated that fewer people watched the highly publicized final episode of *Seinfeld* in 1998 (first aired in 1990; see Seinfeld, Jerry) than watched the final episode of *MASH* in 1983 (first aired in 1972). The trial of former football star O. J. Simpson in 1994 for the murder of his wife (he was acquitted) further demonstrated the hold that cable networks had on American audiences. Some stations carried almost every minute of the lengthy trial live and then filled the evening with talk shows dissecting that day's developments. The effects of television on children, particularly through its emphasis on violence and sex, has long been an issue for social scientists, parents, and politicians (see children's television). In the late 1980s and 1990s, with increased competition brought on by the proliferation of cable networks, talk shows and "tabloid" news shows seemed to broaden further frank or sensational on-air discussion of sex.

In response to government pressure, the television industry decided to display ratings of its programs in 1996. The ratings were designed to indicate the age groups for which the programs might be suitable: TV-G (for general audiences), TV-PG (parental guidance suggested), TV-14 (unsuitable for children under 14), and TV-MA (for mature audiences only). In response to additional complaints, all the networks except NBC agreed the next year to add V (for violence), S (for sex), L (for course language) and D (for suggestive dialogue) to those ratings. Also, the "V-chip" imbedded in new television sets, in accordance with a provision of a telecommunications bill passed in 1996, gave parents the power to automatically prevent their children from watching television programs with inappropriate ratings.

Critics of the ratings saw them as a step toward censorship and questioned whether a TV-14 rating would make a program seem more, not less, attractive to an inquisitive child. In 1997 the federal government gave each U.S. television broadcaster an additional channel on which to introduce high definition television, or HDTV. Initial transmissions of this high-resolution form of television, in which images appear much sharper and clearer, began in 1998. Standard television sets cannot pick up HDTV and will presumably have to be replaced or modified by 2006, when traditional, low-definition television broadcasts are scheduled to end and broadcasters are scheduled to return their original, non-HDTV channel to the government. The HDTV format approved in the United States calls for television signals to be transmitted digitally. This will allow for further convergence between computers, the Internet, and television.

In 1998 it was already possible to view video on the World Wide Web and to see and search television broadcasts on a computer. As computers become more powerful, they should be able to handle video as easily as they now handle text. The television schedule may eventually be replaced by a system in which viewers are able to watch digitally stored and distributed programs or segments of programs whenever they want. Such technological changes, including the spread of new cable networks, have been arriving slower in most other countries than in the United States. Indeed, according to one survey, it was only in the 1990s that the spread of television transmitters, television sets, and electricity made it possible for half of the individuals in the world to watch television. However, television's attraction globally is strong. Those human beings who have a television set watch it, by one estimate, for an average of two-and-a-half hours a day.

Changes In TV Viewing Behavior During the 1970s Through the 1990s

TV continued to serve as a significant group medium through the 1960s and into the 1970s. Although the number of channels available to the average household grew to 5 or 6 in the early 1970s (up from 2 or 3 in the early 1950s), there was relatively little channel changing and the three broadcast networks in the U.S. accounted for 70 to 90 percent of viewing. However, the seeds of change for TV viewing patterns were planted by a series of technological innovations.

First, the average TV screen size increased from 14 inches to 17, 19 and then 21 inch screens by the 1970s. This allowed people to sit farther back from the screen. Second, multiple TV sets began to appear in many households. In 1955, only three percent of U.S. households had more than one TV. By the 1975, this number increased to 40 percent of households and by the mid 1990s, more than 70 percent of U.S. households had two or more TVs. Also, in the 1970s and early 1980s, cable TV systems were introduced into many large cities and the first cable-originated channels appeared such as HBO and Ted Turner's Superstation. This increased the number of channels available to viewers, rising to 45 channels for the average household in 1995.

Among the most significant technological innovations introduced during this period was the remote control. While remotes were available in the 1950s, they did not enter a significant number of households until the 1970s. They then took off and were in nearly every household by 1990. With multiple sets, remote controls and a larger number of channels available, TV viewing became more personalized and channel changing became more common. One individual could now separate from other household members and watch programs with more niche appeal such as sports, from a bedroom, den or basement "TV room."

However, it is important to understand that this was an evolution, not a revolution in viewing behavior. The changes occurred gradually from the early 1970s through the 1990s. Group TV viewing did not disappear. It simply became one of many TV viewing patterns. Further, group viewing itself evolved. Now, a family sitting together in the living room watching TV during the evening was just one form of group viewing. Families also watched as a group in the kitchen or dining room while eating a meal and smaller groups gathered together to watch TV, e.g., a group of children watching from the floor in a den or bedroom. In addition, the number of channels people actually watched grew much more slowly than the number available to them. With 45 channels available in 1995, the average household watched only seven channels in a given week.

During the 1970s and 1980s, the use of televisions changed in another significant way. The TV set became a display monitor for VCR and videogame content as well as a way to watch regular television. With this, multiple boxes were connected to TVs and multiple remotes occupied coffee tables, couches and floors. One impact was the movement of family photos from atop the TV to the side in order to make way for various boxes that now sat on top of the TV. Also, remote controls became a problem as well as a helpful tool for TV viewers. Remotes were sometimes lost in the pile of gear associated with TVs and viewers encountered "mode" problems, i.e., which mode was the TV set for (TV, Video, Auxiliary Device, etc.) and what remote should they use to control which box or function?

The VCR is also instructive because its use changed significantly over time. In the first several years of VCR use in households, time shift recording was the dominant use. In the early 1980s, the average household with a VCR recorded five hours of television per week. Over time, videocassette rentals and purchases surpassed time shift recording as the major use for VCRs. Twenty five years after VCRs were introduced, only one in five U.S. households did any recording at all with their VCR.

By the 1990s, TV viewing encompassed a complex and elaborate set of behaviors. Traditional TV watching that was still the dominant use of television. However, people also used TVs to display many different video services (e.g., videocassettes and videogames) in many different rooms, with different mixes of viewers. TV was on in the average U.S. household for seven hours a day. It was both watched and listened to - many listened to TV as they got dressed, prepared dinner or did housework. In this way, TV took on some of the functions served by radio in earlier decades. It filled the house with sound and pictures and created an environment within which people lived. TV was also a multi-tasking device, as many people opened mail, read newspapers, talked on the phone etc., while watching or listening to TV.

The size of TVs continued to grow in the 1990s. The average "main" TV set was 27 inches and one in four households had a TV that was 30 inches or larger. The "home entertainment center" also emerged. Approximately one in six households had a giant size screen (40 inches or larger), stereo speakers and multiple boxes attached to the TV. In my visits to homes with home entertainment centers it was not unusual to find six, seven or even eight remote controls on a table in front the large-screen TV, which conveyed a feeling of an altar or shrine in a church. It both dominated and defined the room.

The picture of TV viewing in the 1990s is even more complex. While the screen size for the "main" household TV increased, other household TVs varied significantly in screen size. The screen size of these other TVs varied in relation to the room where they were placed and how they were used. Smaller TVs were located in kitchens, bedrooms and even garages. Often, these were placed inside other furniture such as an armoire that dictated the size of the TV as well as its orientation within a room - it could be seen from parts of the room and heard from other parts, depending on where a person stood or sat.

In the 1990s, TV also re-emerged as a significant device in public locations - offices, building lobbies and airport waiting areas. TV watching in public locations was not the main source of TV, as it was for some people in the late 1940s and early 1950s, but a complement to household TV viewing. TVs in public locations may also signal just how dependent people had become on television that they wanted it outside as well as inside the home.

One way to perceive the changes in TV viewing from 1950 to the end of the 1990s, is to think of TV in 1950 as a fireplace or hearth. Families gathered around it and were mesmerized by a magic box. By the 1990s, TV had become an environment that surrounded people no matter where they were in the household and, often, outside the home. People consumed TV constantly but for different uses depending upon the time of day or location where they watched it.

TV VIEWING AND NEW MEDIA

There are many general issues associated with TV viewing and new media. The first relates to acquisition. How and why do people get new media? It is important to distinguish between technologies that people acquire because of what the device can do (e.g., a PVR) and technology that people acquire because of the content it provides (e.g., ITV or an electronic program guide). As obvious at it may seem, it is difficult to get people to acquire a technology that requires content when there is little or no content available (this problem plagued some ITV services in the 1990s). The price of technology also matters, but there are often significant differences between what early adopters will pay and what the mass market will pay. Early adopters are important because they are often willing to pay higher initial prices for a technology and this helps to bring the price down for later adopters. However, there are examples of technologies that could probably have attracted a mass market at a lower price but for which there was no early adopter group willing to pay the high initial cost of the technology (teletext in the U.S. is one example).

Standards and multiple versions of a technology are important as well, but from a consumer perspective the issue is not related to regulation or competition but simplicity and a feeling of comfort that the technology they buy will be around in three or four years. New technologies must also compete for space in the household. This is a more significant issue today in technology-dense households than it was in the 1950s. In my research in households, many people raise the question: where am I going to put another electronic box? Some technologies also face the challenge of whether people can understand what they do. This has been an issue for PVRs. Many people don't know what they do and some marketing materials make them sound like high-end VCRs. Word-of-mouth is also very important in spreading the news about new technologies. However, word-of-mouth is much more valuable when technology adoption reaches a certain threshold: five thousand people spreading good word-of-mouth about a technology is not as valuable as five million spreading good word-of-mouth. This too appears to have been a barrier for PVRs which have received extraordinary word-of-mouth in the US, but from a relatively modest base of users.

A second general issue is how people use a new medium when they acquire it and, specifically, whether they change existing habits. In my research, most new technologies and services enter a household with fixed media habits. Technology is in certain rooms; people sit in a favorite chair while using a medium; and schedules dictate when they use media. Initially, people make very few changes to their existing patterns of media use to accommodate new technology such as sitting closer to a screen that has menu items in small text. This has been the case in my studies of ITV. Initially, people changed little in terms of where they sat or when they used ITV compared with regular television. Changes in behavior do occur but generally they evolve slowly. PVRs are an exception. They introduced a number of changes in TV use very quickly for those who adopted the technology.

Behavior When Using TV and Broadband PCs, PVRs, EPGs and ITV

We are in the middle of an exciting time when a number of technologies and services are being introduced into households. Their long term impacts on TV viewing are hard to predict. However, some early hints of what is to come are emerging. Below, I review briefly some of the TV-related behavior associated with four of these new technologies, based upon my own research and other published studies.

TV and Broadband PCs In the late 1990s, a feature article in the Sunday *New York Times Magazine* declared that interactive *television* was dead and, in its place, the broadband Web would become the sole delivery vehicle for interactive TV. Broadband has yet to deliver interactive streaming video in a significant way. People who experience video on PCs are more likely to do so through high-end video games. The PC in some homes has also become a secondary DVD player for movies, used mostly by children when the primary DVD player in the household is occupied by someone else.

PCs with broadband connections in households are a very interesting sociological phenomenon. Typically, they are on whenever anyone is in the house and people use them for much longer periods compared to dial-up PCs. In some cases, people sit in easy chairs at these PCs instead of office chairs, perhaps because they spend so much time with the PC. Also, many decorate the area on top of or next to the monitor with stuffed animals or family photos, much like they decorated TVs in the 1960s and beyond. Also, I have been in a number of homes where a broadband PC is located in the "computer room" - a term some people have adopted to signify that the PC defines the use of the room, much as some people began to use the term "television room" a few decades ago.

Rather than replace TV, broadband PCs are increasingly likely to be in the same room as a television and to be used with a TV. There are eight different patterns of use for TVs and PCs in the same room that I have observed and many more that I probably have not yet seen. The first clue to how the two devices are used together is where each is placed in the room - side by side, on opposite walls or spaced widely apart on the same wall. In some cases, people use a PC and television at the same time, e.g., viewing the Web site of a channel that they are currently watching on TV. Some game shows promote this as a form of interactive television. In other cases, people surf the Web during commercials or send instant messages to friends while watching TV (the instant message may be about the show they are watching or other topics). In addition, some people surf the Web and listen to TV, turning around to watch when something interesting catches their ear. In other cases, one person uses the PC while others watch TV in the same room. The use of a PC and TV in the same room may also vary by time of day, with the broadband PC receiving more usage during the day (e.g., in a home office) and TV receiving more use during the evening. These multiple patterns of use are likely to evolve over time and may be reduced to a smaller set of uses for the two media. However, there are no signs yet that the two boxes will merge, as some have predicted.

Personal Video Recorders (PVRs) PVRs have had an extraordinary impact on TV viewing behavior of those who have the technology. People with PVRs quickly move away from real time viewing of scheduled programs and either watch programs that have been stored on the hard drive or turn on the PVR and start watching programs 10 to 15 minutes later so that they can skip ahead if they wish. Many PVR owners call the latter, "building a buffer," and use the time delay to skip past boring scenes in programs or commercials. They also report that television is a more enjoyable experience because they watch more of the programs they like rather than settle for what is on at any given time. Curiously, the group that seems to benefit most from PVRs are those who have limited time to watch TV and whose schedules are inflexible. They report that the full 24 hour schedule is now available to them even though they may be able to watch TV for only one hour a night. This also appears to have a modest benefit for programs that are not in prime time and therefore not available to the largest audience. However, published reports on PVR usage indicate that people tend to record the most popular shows.

There are many early indicators of side effects associated with PVR usage. One is that PVR owners channel surf less, since they have already found what they want from the full week's schedule that is now available to them. There also appears to be slightly more group viewing in PVR homes, since it is on one TV and many in the household want to get the benefits of the PVR. The impact on commercials has been discussed at great length - people with PVRs watch fewer commercials. However, there are several important nuances to this behavior. Many people report that they have stopped watching commercials completely, but follow-up discussion with these people indicates that they still watch some commercials, e.g., promotions for TV programming, funny 'favorite' commercials and commercials with visuals that catch their eye as they fast forward through the commercial. Indeed, it was reported that during the last Super Bowl, the most common use of the instant replay on PVRs was to watch commercials a second time.

Given the very positive reaction to PVRs by those who have purchased them, it is a puzzle why they have been adopted so slowly (approximately one million users in the U.S. after three years in the marketplace). The large size of the first generation of PVR boxes, $10 per month subscription fees and poor understanding by consumers about what PVRs do may be contributing factors. In the long term, if PVRs are widely adopted, the impact on TV viewing could be dramatic, freeing viewers from program schedules except for live events and impacting how advertising is delivered.

Electronic Program Guides (EPGs) EPGs are 'quiet' new services compared to PVRs and broadband video, yet they have had a broad impact to date both in terms of adoption and TV viewing. In the U.S., EPGs are provided within digital cable or satellite services, or are part of a larger ITV package, and reach approximately 30 million viewers. They replaced older 'scrolling' TV guides that do not allow viewers to control the time period or channels that are displayed. People who have EPGs like them, use them increasingly over time and say that they would not give them up. However, EPGs have received less press attention compared to their more exciting interactive cousins such as PVRs and full ITV packages.

EPGs are used most frequently at the beginning of a session of TV watching, during program breaks and during commercials. EPGs differ somewhat in design. One feature that is part of some EPGs and which viewers like is a TV window within the Guide that displays whatever channel (video and sound) a person was watching before they entered the Guide. This lets people go into the EPG during commercials or the break in action of a sporting event and return when regular programming or the next play in the sporting event starts. Curiously, this has the side effect of exposing people to commercials (even though within a window on the TV screen) that they might otherwise have missed. Many EPG users report that previously they would channel surf during commercials but now they go into the Guide instead. EPG users also report less channel surfing generally. They now find channels to watch from the Guide rather than by channel surfing.

Most EPG behavior involves searching for programming information in the next hour or a couple of hours later in the day. Advanced searches and browsing through program schedules days ahead are used much less. Most EPGs have a related feature, sometimes called Info or Display, which brings up information about a specific program in the bottom or top third of the screen, superimposed over the regular video of a program. This feature is also used a great deal in EPG homes, often as much as the main Guide. Viewers report that they bring up the Info or Display feature when they first turn on a TV set, to "anchor" them and tell them where they are as well as the content on that channel. Later, many use Info or Display when channel surfing. It provides more information than they are likely to get by watching a program for two or three seconds.

In the long term, EPGs could become a mechanism for steering viewers to certain programs or channels, based upon the prominence given to those programs or channels within the EPG screen "real estate." In addition, they appear to be "organizing" the way viewers group channels in their minds. The more people use EPGs, the more the EPG model of channels and programs is likely to become the model for how people think about television.

Interactive Television (ITV) ITV includes a broad range of new services that provide video-on-demand, interactive games over TV, polling, interactive program guides (discussed above), multiple camera angles in sporting events, and various new uses of text or graphics over a TV picture, among other ways to interact with the TV - all under the control of a viewer. Europe is ahead of the U.S. in deploying interactive television. Nonetheless, U.S. experiences can contribute to our overall understanding of the appeal and use of ITV.

There does not appear to be a single driving feature that attracts people to ITV. Indeed, some people who get ITV are not sure what it is or what it can provide them. One study in Europe reported that many people with ITV did not realize that they had it. The word "interactive" may itself be a problem in conveying what these services offer. The words for component services within ITV such as video-on-demand, program guide and games seem to be more readily understood. Initially, viewers do not change where they sit while using ITV, although in some cases it would be better if they sat closer so that they could more easily read menu screens. This potential problem is mitigated at the moment by the fact that ITV is not typically on all household sets. It is generally installed on the 'main' household TV that is larger than other TVs in the home. Further, people tend to sit in chairs oriented towards the main TV and watch the screen more than they do on some other household TVs, which are smaller and may be in a corner of the room. Placement of ITV on the main household TV also encourages more group use of ITV services. Existing unwritten rules in the household about who controls the remote control do not change when it is an ITV remote. This means that males are more likely to control the ITV remote. However, this varies in relation to the program and time of day.

Expectations about ITV are another important issue, especially when Web-related or actual Web screens are offered as part of an ITV service. When screen design or wording in an ITV menu give people a 'feeling' that it is like the Web, viewer expectations are raised. Some want everything that is on the Web, only in an interactive video format. Service providers have tried to control expectations with a "walled garden" approach that provides a limited range of services but and maintains quality within that range of services.

What do people like and use within ITV? First, most ITV services provide digital signals and extra channels. People have responded well to better picture quality and more channels. EPGs are also liked and used. Video-on-demand is appealing to those who have experienced it. The single most attractive feature of VOD seems to be Pause that lets a viewer stop a movie and answer the telephone or go to the bathroom. Not having to return a rented videocassette has also been very attractive. However, ITV user expectations for video-on-demand content are high. They expect the same movies that are available at the video rental shops and this has not been the case so far, limiting the use and appeal of current VOD services.

An alternative form of video-on-demand, Subscription Video-on-Demand (SVOD) is also being tested. With SVOD, a viewer can access all the movies available on HBO or Starz in a given month - at any time and with full VCR controls. SVOD inherently meets viewer expectations about content that will be available, since it doesn't promise to deliver any content other than that which is on the regular schedule of the pay movie channel for that month. The perceived benefit of SVOD is convenience. A viewer does not have to fit their schedule into the timeframe of the movie service provider. Other ITV content such as interactive shopping, games and multiple camera angles of sporting events have been limited so far, at least in the U.S. context. Also, there are no betting services and limited adult content. If history is a guide, the latter two services will be appealing. In addition, some ITV services have relied primarily on advertising to attract viewers. That is, an ITV user can click on a commercial and see more information. This certainly has a place in an ITV service with entertainment and information content as well as advertising. However, it stretches the imagination to think that viewers will pay for an ITV service whose primary content offer is advertising.

Navigating through ITV menus and finding services is a not-so-surprising challenge. In the past, television was a linear medium in terms of channel organization. Channels started at 1 and went up to 12, 40, or 200. If a new channel was introduced, sooner or later a TV viewer was likely to stumble it. With many ITV services, a viewer accesses choices from a tree-and-branch structure menu or a key on the remote control. Since there is limited screen space on a main menu, a person may have to make three or four choices before reaching some services. Further, since there are so many remotes in households, each with dozens of keys, a dedicated ITV key could become lost in the forrest of keys. Designing easy-to-use menus and a new gewneration of remote controls is a challenge.

The words people use in describing ITV are informative. Curiously, they do not use the word "interactive" as much as words such as "control" and "convenience." That is, ITV gives them a feel of control over television watching and it fits conveniently into their schedule rather than requiring them to follow a fixed television schedule, as in the past. Another curious word that crops up when people talk about their ITV experience is "go." That is, when talking about regular TV channels, they say, "I watch HBO" or "I watch NBC." However, when talking about ITV, many say, "I go to the Guide" or "I go and order a movie." Apparently, the ITV experience gives them a feeling that ITV services are *places* they go to. Some teenagers even say, "I hang out at (an ITV service)."

Discussion

This analysis of the evolution of TV viewing raises a number of questions about the future. Among the questions for discussion and further research are:

- Do people really want to interact with television? I believe that the answer is clearly yes. However, some people want to interact a lot and some want to interact only a little. Further, the amount of interaction is likely to vary by type of program (higher for a game show, lower for a situation comedy) and even time of day. ITV can serve high and low levels of interactivity. For example, in a VOD service, a person interacts with a menu of movie listings, selects one and then sits back for two hours to watch the movie, perhaps pausing at some point to get food from the kitchen.
- Are TV schedules going to go away, except for live events? It could happen, but the timing is unclear. One way to think about PVRs is as a video server for the home. VOD provides a video server at the cable headend. If both are accepted within the marketplace, TV schedules could become a part of the history of TV viewing. If this happens, it is likely to take a decade or more. However, what are the implications for TV viewing as well as program production and distribution in such an environment?
- What is the best way to organize and access a TV channel or ITV service? Current options include a traditional linear channel lineup accessed by up/down arrows on a remote or direct entry; an on-screen menu; or special keys on a remote control. The linear channel lineup becomes problematic with hundreds of channels/services;

special keys on a remote are problematic in an environment where a household has several remotes, each with dozens of keys; and menus require sub-menus and raise the possibility that some people will become lost while trying to find channels or ITV services. The design challenge is significant.

- What is the continuing role of public locations in educating or exposing the public to new television services? In the U.S., TV set manufacturers, broadcasters and electronic stores have a presumed interest in promoting HDTV. Yet, it is amazing how few public locations exist where people can experience HDTV. ITV, PVRs and other new services are also generally not available in public locations.
- How will the design of traditional linear television programs change as viewers gain experience with ITV, PVRs and the broadband Web? In the U.S., some MTV programs look like a Web site or ITV, with multiple video windows, although the viewer does not control the video. CNBC, CNN, Fox News, and Bloomberg also use multiple video windows. Some producers dislike these formats, arguing that they disrupt the integrity of core content. However, the formats feed the appetites of people who seek many inputs at once and have become accustomed to receiving them on the Web or through ITV.
- What is the future of advertising in both traditional television, ITV and TV watched through PVRs? If people channel change during commercials and fast forward through them on a PVR, advertising and television may change in any of several ways: 1/ the viewer could pay for more content and there would be less advertising; 2/ sponsorships of programs, common in 1950s commercial television and used today on public television in the U.S., could return; 3/ product placements within programs could increase; 4/ more advertising could be digitally inserted as backgrounds within content; 5/ a "newspaper" model of advertising could be employed on television, in which content and ads appear on the same screen, each in different windows; 6/ icons or message "flags" could be inserted in commercials to signal PVR users about the type of commercial they are fast forwarding through; and, 7/ prime shelf space in EPGs and ITV menus may become valuable advertising venues.
- Will television services have a home page in the future? Some current ITV services have a home page, much like television services in hotels. If television home pages become the norm, they could potentially give a lot of power to the organizations that control those home pages.
- Is ITV on the Web dead? Just as it was premature for the *New York Times* to declare the death of interactive *television* in the late 1990s

and predict that the future of ITV would be on the broadband Web, it is premature now to say that the Web will never be a medium for ITV. What is the likely role for interactive video on the Web in the future? When will it be feasible and how are people likely to use video on the broadband Web?

- Will ITV and other advanced television services change the core functions of TV? TV viewing will continue to evolve. However, it is unlikely that the core functions of television - to entertain, to inform and to provide relaxation from the hectic world of everyday life - will change. This is important to remember as we create content and are tempted by high-tech wizardry to move away from the simple but very important functions that TV serves in people's lives.

Index